Being There

Xiii

Carl Bly
St. Paul
FEBRUARY 1999

Being There

Putting Brain, Body, and World Together Again

Andy Clark

A Bradford Book
The MIT Press
Cambridge, Massachusetts
London, England

Second printing, 1997

Set in Sabon by The MIT Press.
Printed and bound in the United States of America.

Library of Congress Cataloging-in-Publication Data

Clark, Andy.
Being there : putting brain, body, and world together again / Andy Clark.
p. cm.
"A Bradford book."
Includes bibliographical references (p.) and index.
ISBN 0-262-03240-6
1. Philosophy of mind. 2. Mind and body. 3. Cognitive science. 4. Artificial intelligence. I. Title.
BD418.3.C53 1996
153—dc20 96-11817
CIP

Cover illustration: *Starcatcher* (1956) by Remedios Varo. The author wishes to thank Walter Gruen for his generosity in granting permission.

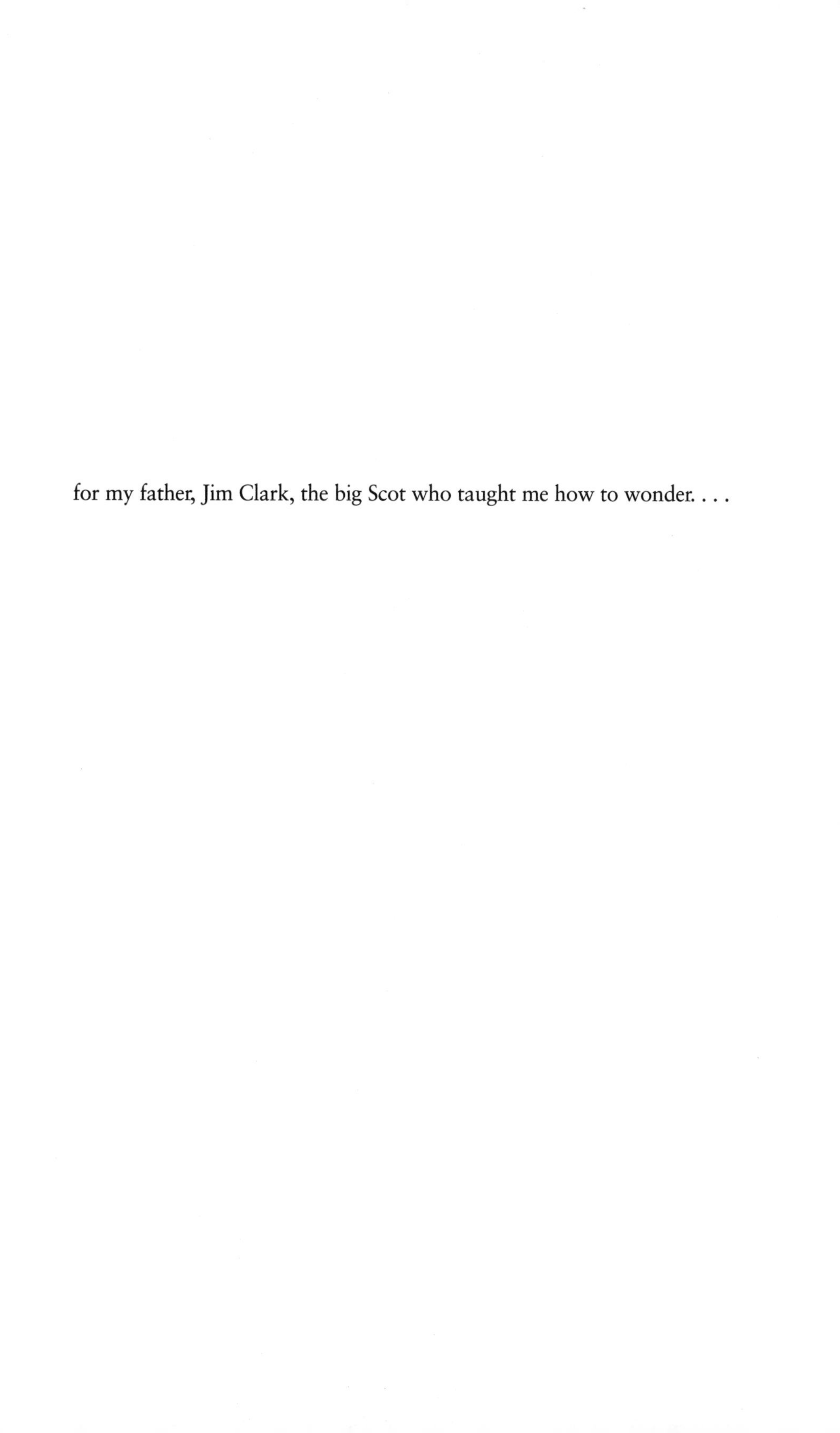

for my father, Jim Clark, the big Scot who taught me how to wonder. . . .

Contents

Preface: Deep Thought Meets Fluent Action

If you had to build an intelligent agent, where would you begin? What strikes you as the special something that separates the unthinking world of rocks, waterfalls, and volcanos from the realms of responsive intelligence? What is it that allows some parts of the natural order to survive by perceiving and acting while the rest stay on the sidelines, thought-free and inert?

"Mind," "intellect," "ideas": these are the things that make the difference. But how should they be understood? Such words conjure nebulous realms. We talk of "pure intellect," and we describe the savant as "lost in thought." All too soon we are seduced by Descartes' vision: a vision of mind as a realm quite distinct from body and world.[1] A realm whose essence owes nothing to the accidents of body and surroundings. The (in)famous "Ghost in the Machine."[2]

Such extreme opposition between matter and mind has long since been abandoned. In its stead we find a loose coalition of sciences of the mind whose common goal is to understand how thought itself is materially possible. The coalition goes by the name *cognitive science*, and for more than thirty years computer models of the mind have been among its major tools. Theorizing on the cusp between science fiction and hard engineering, workers in the subfield known as *artificial intelligence*[3] have tried to give computational flesh to ideas about how the mind may arise out of the workings of a physical machine—in our case, the brain. As Aaron Sloman once put it, "Every intelligent ghost must contain a machine."[4] The human brain, it seems, is the mechanistic underpinning of the human mind. When evolution threw up complex brains, mobile bodies, and nervous systems, it opened the door (by purely physical means) to whole new ways of living and adapting—ways that place us on one side of a natural divide, leaving volcanos, waterfalls, and the rest of cognitively inert creation on the other.

But, for all that, a version of the old opposition between matter and mind persists. It persists in the way we study brain and mind, excluding as "peripheral" the roles of the rest of the body and the local environment. It persists in the tradition of modeling intelligence as the production of symbolically coded solutions to symbolically expressed puzzles. It persists in the lack of attention to the ways the body and local environment are literally built into the processing loops that result in intelligent action. And it persists in the choice of problem domains: for example, we model chess playing by programs such as Deep Thought[5] when we still can't get a real robot to successfully navigate a crowded room and we still can't fully model the adaptive success of a cockroach.

In the natural context of body and world, the ways brains solve problems is fundamentally transformed. This is not a deep philosophical fact (though it has profound consequences). It is a matter of practicality. Jim Nevins, who works on computer-controlled assembly, cites a nice example. Faced with the problem of how to get a computer-controlled machine to assemble tight-fitting components, one solution is to exploit multiple feedback loops. These could tell the computer if it has failed to find a fit and allow it to try to again in a slightly different orientation. This is, if you like, the solution by Pure Thought. The solution by Embodied Thought is quite different. Just mount the assembler arms on rubber joints, allowing them to give along two spatial axes. Once this is done, the computer can dispense with the fine-grained feedback loops, as the parts "jiggle and slide into place as if millions of tiny feedback adjustments to a rigid system were being continuously computed."[6] This makes the crucial point that treating cognition as pure problem solving invites us to abstract away from the very body and the very world in which our brains evolved to guide us.

Might it not be more fruitful to think of brains as controllers for embodied activity? That small shift in perspective has large implications for how we construct a science of the mind. It demands, in fact, a sweeping reform in our whole way of thinking about intelligent behavior. It requires us to abandon the idea (common since Descartes) of the mental as a realm distinct from the realm of the body; to abandon the idea of neat dividing lines between perception, cognition, and action[7]; to abandon the idea of an executive center where the brain carries out high-level reasoning[8]; and most of all, to abandon research methods that artificially divorce thought from embodied action-taking.

What emerges is nothing less than a new science of the mind: a science that, to be sure, builds on the fruits of three decades' cooperative research, but a science whose tools and models are surprisingly different—a cognitive science of the embodied mind. This book is a testimony to that science. It traces some of its origins, displays its flavor, and confronts some of its problems. It is surely not the last new science of mind. But it is one more step along that most fascinating of journeys: the mind's quest to know itself and its place in nature.

Acknowledgments

Parts of chapters 6 and 9 and the epilogue are based on the following articles of mine. Thanks to the editors and the publishers for permission to use this material.

"Happy couplings: Emergence, explanatory styles and embodied, embedded cognition," in *Readings in the Philosophy of Artificial Life*, ed. M. Boden. Oxford University Press.

"Economic reason: The interplay of individual learning and external structure," in *Frontiers of Institutional Economics*, ed. J. Drobak. Academic Press.

"I am John's brain," *Journal of Consciousness Studies* 2 (1995), no. 2: 144–148.

Sources of figures are credited in the legends.

Groundings

Being There didn't come from nowhere. The image of mind as inextricably interwoven with body, world, and action, already visible in Martin Heidegger's *Being and Time* (1927), found clear expression in Maurice Merleau-Ponty's *Structure of Behavior* (1942). Some of the central themes are present in the work of the Soviet psychologists, especially Lev Vygotsky; others owe much to Jean Piaget's work on the role of action in cognitive development. In the literature of cognitive science, important and influential previous discussions include Maturana and Varela 1987, Winograd and Flores 1986, and, especially, *The Embodied Mind* (Varela et al. 1991). *The Embodied Mind* is among the immediate roots of several of the trends identified and pursued in the present treatment.

My own exposure to these trends began, I suspect, with Hubert Dreyfus's 1979 opus *What Computers Can't Do*. Dreyfus's persistent haunting of classical artificial intelligence helped to motivate my own explorations of alternative computational models (the connectionist or parallel distributed processing approaches; see Clark 1989 and Clark 1993) and to cement my interest in biologically plausible images of mind and cognition. Back in 1987 I tested these waters with a short paper, also (and not coincidentally) entitled "Being There," in which embodied, environmentally embedded cognition was the explicit topic of discussion. Since then, connectionism, neuroscience, and real-world robotics have all made enormous strides. And it is here, especially in the explosion of research in robotics and so-called artificial life (see e.g. papers in Brooks and Maes 1994), that we finally locate the most immediate impetus of the present discussion. At last (it seems to me), a more rounded, compelling, and integrative picture is emerging—one that draws together many of the elements of the previous discussions, and that does so in a framework rich in practical illustrations and concrete examples. It is this larger, more integrative picture that I here set out to display and examine.

The position I develop owes a lot to several authors and friends. At the top of the list, without a doubt, are Paul Churchland and Dan Dennett, whose careful yet imaginative reconstructions of mind and cognition have been the constant inspiration behind all my work. More recently, I have learned a lot from interactions and exchanges with the roboticists Rodney Brooks, Randall Beer, Tim Smithers, and John Hallam. I have also been informed, excited, and challenged by various fans of dynamic systems theory, in particular Tim van Gelder, Linda Smith, Esther Thelen, and Michael Wheeler. Several members of the Sussex University Evolutionary Robotics Group have likewise been inspiring, infuriating, and always fascinating—especially Dave Cliff and Inman Harvey.

Very special thanks are due to Bill Bechtel, Morten Christiansen, David Chalmers, Keith Butler, Rick Grush, Tim Lane, Pete Mandik, Rob Stufflebeam, and all my friends, colleagues, and students in the Philosophy/Neuroscience/Psychology (PNP) program at Washington University in St. Louis. It was there, also, that I had the good fortune to encounter Dave Hilditch, whose patient attempts to integrate the visions of Merleau-Ponty and contemporary cognitive science were a source of joy and inspiration. Thanks too to Roger Gibson, Larry May, Marilyn Friedman, Mark Rollins, and all the members of the Washington University Philosophy Department for invaluable help, support, and criticism.

David van Essen, Charlie Anderson, and Tom Thach, of the Washington University Medical School deserve special credit for exposing me to the workings of real neuroscience—but here, especially, the receipt of thanks should not exact any burden of blame for residual errors or misconceptions. Doug North, Art Denzau, Norman Schofield, and John Drobak did much to smooth and encourage the brief foray into economic theory that surfaces in chapter 9—thanks too to the members of the Hoover Institute Seminar on Collective Choice at Stanford University. I shouldn't forget my cat, Lolo, who kept things in perspective by sitting on many versions of the manuscript, or the Santa Fe Institute, which provided research time and critical feedback at some crucial junctures—thanks especially to David Lane, Brian Arthur, Chris Langton, and Melanie Mitchell for making my various stays at the Institute such productive ones. Thanks also to Paul Bethge, Jerry Weinstein, Betty Stanton, and all the other folks at The MIT Press—your support, advice, and enthusiasm helped in so many ways. Beth Stufflebeam provided fantastic help throughout the preparation of the manuscript. And Josefa Toribio, my wife and colleague, was critical, supportive, and inspiring in perfect measure. My heartfelt thanks to you all.

Being There

Introduction: A Car with a Cockroach Brain

Where are the artificial minds promised by 1950s science fiction and 1960s science journalism? Why are even the best of our "intelligent" artifacts still so unspeakably, terminally dumb? One possibility is that we simply misconstrued the nature of intelligence itself. We imagined mind as a kind of logical reasoning device coupled with a store of explicit data—a kind of combination logic machine and filing cabinet. In so doing, we ignored the fact that minds evolved to make things happen. We ignored the fact that the biological mind is, first and foremost, an organ for controlling the biological body. Minds make motions, and they must make them fast—before the predator catches you, or before your prey gets away from you. Minds are *not* disembodied logical reasoning devices.

This simple shift in perspective has spawned some of the most exciting and groundbreaking work in the contemporary study of mind. Research in "neural network" styles of computational modeling has begun to develop a radically different vision of the computational structure of mind. Research in cognitive neuroscience has begun to unearth the often-surprising ways in which real brains use their resources of neurons and synapses to solve problems. And a growing wave of work on simple, real-world robotics (for example, getting a robot cockroach to walk, seek food, and avoid dangers) is teaching us how biological creatures might achieve the kinds of fast, fluent real-world action that are necessary to survival. Where these researches converge we glimpse a new vision of the nature of biological cognition: a vision that puts explicit data storage and logical manipulation in its place as, at most, a secondary adjunct to the kinds of dynamics and complex response loops that couple real brains, bodies, and environments. Wild cognition, it seems, has (literally) no time for the filing cabinet.

Of course, not everyone agrees. An extreme example of the opposite view is a recent $50 million attempt to instill commonsense understanding in a computer by giving it a vast store of explicit knowledge. The project, known as CYC (short for "encyclopedia"), aims to handcraft a vast knowledge base encompassing a significant fraction of the general knowledge that an adult human commands. Begun in 1984, CYC aimed at encoding close to a million items of knowledge by 1994. The project was to consume about two person-centuries of data-entry time. CYC was supposed, at the end of this time, to "cross over": to reach a point where it could directly read and assimilate written texts and hence "self-program" the remainder of its knowledge base.

The most noteworthy feature of the CYC project, from my point of view, is its extreme faith in the power of explicit symbolic representation: its faith in the internalization of structures built in the image of strings of words in a public language. The CYC representation language encodes information in units ("frames") such as the following:

Missouri
Capital: (Jefferson City)
Residents: (Andy, Pepa, Beth)
State of: (United States of America)

The example is simplified, but the basic structure is always the same. The unit has "slots" (the three subheadings above), and each slot has as its value a list of entities. Slots can reference other units (for example, the "residents" slot can act as a pointer to another unit containing still more information, and so on and so on). This apparatus of units and slots is augmented by a more powerful language (the CycL Constraint language) that allows the expression of more complex logical relationships, such as "For all items, if the item is an X then it has property Y." Reasoning in CYC can also exploit any of several simple inference types. The basic idea, however, is to let the encoded knowledge do almost all the work, and to keep inference and control structure simple and within the bounds of current technology. CYC's creators, Douglas Lenat and Edward Feigenbaum

(1992, p. 192), argue that the bottleneck for adaptive intelligence is *knowledge*, not inference or control.

The CYC knowledge base attempts to make explicit all the little things we know about our world but usually wouldn't bother to say. CYC thus aims to encode items of knowledge we all have but seldom rehearse—items such as the following (ibid., p. 197):

Most cars today are riding on four tires.
If you fall asleep while driving, your car will start to head out of your lane pretty soon.
If something big is between you and the thing you want, you probably will have to go around it.

By explicitly encoding a large fraction of this "consensus reality knowledge," CYC is supposed to reach a level of understanding that will allow it to respond with genuine intelligence. It is even hoped that CYC will use analogical reasoning to deal sensibly with novel situations by finding partial parallels elsewhere in its vast knowledge base.

CYC is an important and ambitious project. The commonsense data base it now encodes will doubtless be of great practical use as a resource for the development of better expert systems. But we should distinguish two possible goals for CYC. One would be to provide the best simulacrum of commonsense understanding possible within a fundamentally unthinking computer system. The other would be to create, courtesy of the CYC knowledge base, the first example of a genuine artificial mind.

Nothing in the performance of CYC to date suggests that the latter is in the cards. CYC looks set to become a bigger, fancier, but still fundamentally brittle and uncomprehending "expert system." Adding more and more knowledge to CYC will not remedy this. The reason is that CYC lacks the most basic kinds of adaptive responses to an environment. This shortcoming has nothing to do with the relative paucity of the knowledge the system explicitly encodes. Rather, it is attributable to the lack of any fluent coupling between the system and a real-world environment posing real-world problems of acting and sensing. Even the lowly cockroach, as we shall see, displays this kind of fluent coupling—it displays a version of the kind of robust, flexible, practical intelligence that most computer systems so profoundly lack. Yet such a simple creature can hardly be accused of commanding a large store of explicitly represented knowledge! Thus, the CYC project, taken as an attempt to create genuine intelligence and understanding in a machine, is absolutely, fundamentally, and fatally flawed. Intelligence and understanding are rooted not in the presence and manipulation of explicit, language-like data structures, but in something more earthy: the tuning of basic responses to a real world that enables an embodied organism to sense, act, and survive.

This diagnosis is not new. Major philosophical critics of AI have long questioned the attempt to induce intelligence by means of disembodied symbol manipulation and have likewise insisted on the importance of situated reasoning (that is, reasoning by embodied beings acting in a real physical environment). But it has been all too easy to attribute such doubts to some sort of residual mysticism—to unscientific faith in a soul-like mental essence, or to a stubborn refusal to allow science to trespass on the philosophers' favorite terrain. But it is now increasingly clear that the alternative to the "disembodied explicit data manipulation" vision of AI is not to retreat from hard science; it is to pursue some even harder science. It is to put intelligence where it belongs: in the coupling of organisms and world that is at the root of daily, fluent action. From CYC to cycle racing: such is the radical turn that characterizes the new sciences of the embodied mind.

Take, for example, the humble cockroach. The roach is heir to a considerable body of cockroach-style commonsense knowledge. At least, that is how it must appear to any theorist who thinks explicit knowledge is the key to sensible-looking real-world behavior! For the roach is a formidable escape artist, capable of taking evasive action that is shaped by a multitude of internal and external factors. Here is a brief list, abstracted from Ritzmann's (1993) detailed study, of the escape skills of the American cockroach, *Periplaneta americana*:

The roach senses the wind disturbance caused by the motion of an attacking predator.
It distinguishes winds caused by predators from normal breezes and air currents.
It does not avoid contact with other roaches.
When it does initiate an escape motion, it does not simply run at random. Instead, it takes into account its own initial orientation, the presence of

obstacles (such as walls and corners), the degree of illumination, and the direction of the wind.

No wonder they always get away! This last nexus of contextual considerations, as Ritzmann points outs, leads to a response that is much more intelligent than the simple "sense predator and initiate random run" reflex that cockroach experts (for such there be) once imagined was the whole story. The additional complexity is nicely captured in Ritzmann's descriptions of a comparably "intelligent" automobile. Such a car would sense approaching vehicles, but it would ignore those moving in normal ways. If it detected an impeding collision, it would automatically initiate a turn that took its own current state (various engine and acceleration parameters) into account, took account of the road's orientation and surface, and avoided turning into other dangers. A car with the intelligence of a cockroach, it seems clear, would be way ahead of the current state of the automotive art. "Buy the car with the cockroach brain" does not immediately strike you as a winner of an advertising slogan, however. Our prejudice against basic forms of biological intelligence and in favor of bigger and fancier "filing cabinet/logic machines" goes all too deep.

How *does* the roach manage its escapes? The neural mechanisms are now beginning to be understood. Wind fronts are detected by two cerci (antenna-like structures located at the rear of the abdomen). Each cercus is covered with hairs sensitive to wind velocity and direction. Escape motions are activated only if the wind is accelerating at 0.6 m/s^2 or more: this is how the creature discriminates ordinary breezes from the lunges of attackers. The interval between sensing and response is very short: 58 milliseconds for a stationary roach and 14 milliseconds for a walking roach. The initial response is a turn that takes between 20 and 30 milliseconds (Ritzmann 1993, pp. 113–116). The basic neural circuitry underlying the turn involves populations of neurons whose locations and connectivity are now quite well understood. The circuitry involves more than 100 interneurons that act to modulate various turning commands in the light of contextual information concerning the current location of the roach and the state of the local environment. The basic wind information is carried by a population of ventral giant interneurons, but the final activity builds in the results of modulation from many other neuronal populations sensitive to these other contextual features.

Confronted with the cockroach's impressive display of sensible escape routines, a theorist might mistakenly posit some kind of stored quasi-linguistic database. In the spirit of CYC, we might imagine that the roach is accessing knowledge frames that include such items as these:

If you are being attacked, don't run straight into a wall.
If something big is between you and the food, try to go around it.
Gentle breezes are not dangerous.

As the philosopher Hubert Dreyfus (1991) and others have pointed out, the trouble is that real brains don't seem to use such linguaform, text-like resources to encode skillful responses to the world. And this is just as well, since such strategies would require vast amounts of explicit data storage and search and could thus not yield the speedy responses that real action requires. In fact, a little reflection suggests that there would be no obvious end to the "commonsense" knowledge we would have to write down to capture all that an adult human knows. Even the embodied knowledge of a cockroach would probably require several volumes to capture in detail!

But how else might AI proceed? One promising approach involves what has become known as *autonomous-agent theory*. An autonomous agent is a creature capable of survival, action, and motion in real time in a complex and somewhat realistic environment. Many existing artificial autonomous agents are real robots that are capable of insect-style walking and obstacle avoidance. Others are computer simulations of such robots, which can thus move and act only in simulated, computer-based environments. There are disputes between researchers who favor only real-world settings and real robots and researchers who are happy to exploit "mere" simulations, but the two camps concur in stressing the need to model realistic and basic behaviors and in distrusting overintellectualized solutions of the "disembodied explicit reasoning" stripe.

With this general image of autonomous-agent research in mind, let us return very briefly to our hero, the cockroach. Randall Beer and Hillel Chiel have created plausible computer and robot simulations of cockroach locomotion and escape. In modeling the escape response, Beer and Chiel set out to develop an autonomous-agent model highly constrained by ethological and neuroscientific data. The goal was, thus, to stay as close to the real biological data as is currently possible. To this end, they combined the autonomous-agent methodology with neural-network-style modeling. They also constrained this computational model in ways consistent with what is known about the actual neural organization of (in this case) the cockroach. They used a neural net to control the body of a simulated insect (Beer and Chiel 1993). The net circuitry was constrained by known facts about the neural populations and connectivities underlying the escape response in real cockroaches. After training, the neural network controller was able to reproduce in the simulated insect body all the main features of the escape response discussed earlier. In the chapters that follow, we shall try to understand something of how such successes are achieved. We shall see in detail how the types of research just sketched combine with developmental, neuroscientific, and psychological ideas in ways that can illuminate a wide range of both simple and complex behaviors. And we shall probe the surprising variety of adaptive strategies available to embodied and environmentally embedded agents—beings that move and that act upon their worlds.

These introductory comments set out to highlight a fundamental contrast: to conjure the disembodied, atemporal intellectualist vision of mind, and to lay beside it the image of mind as a controller of embodied action. The image of mind as controller forces us to take seriously the issues of time, world, and body. Controllers must generate appropriate actions, rapidly, on the basis of an ongoing interaction between the body and its changing environment. The classical AI planning system can sit back and take its time, eventually yielding a symbolically couched description of a plausible course of action. The embodied planning agent must take action fast—before the action of another agent claims its life. Whether symbolic, text-like encodings have any role to play in these tooth-and-claw decisions is still uncertain, but it now seem clear that they do not lie at its heart.

The route to a full computational understanding of mind is, to borrow a phrase from Lenat and Feigenbaum, blocked by a mattress in the road. For many years, researchers have swerved around the mattress, tried to finesse it away, done just about anything except get down to work to shift it. Lenat and Feigenbaum think the mattress is knowledge—that the puzzles of mind will fall away once a nice big knowledge base, complete with explicit formulations of commonsense wisdom, is in place. The lessons of wild cognition teach us otherwise. The mattress is not knowledge but basic, real-time, real-world responsiveness. The cockroach has a kind of common sense that the best current artificial systems lack—no thanks, surely, to the explicit encodings and logical derivations that may serve us in a few more abstract domains. At root, our minds too are organs for rapidly initiating the next move in real-world situations. They are organs exquisitely geared to the production of actions, laid out in local space and real time. Once mind is cast as a controller of bodily action, layers upon layers of once-received wisdom fall away. The distinction between perception and cognition, the idea of executive control centers in the brain, and a widespread vision of rationality itself are all called into question. Under the hammer too is the methodological device of studying mind and brain with scant regard for the properties of the local environment or the opportunities provided by bodily motion and action. The fundamental shape of the sciences of the mind is in a state of flux. In the chapters to follow, we will roam the landscape of mind in the changing of the light.

I

Outing the Mind

Well, what do you think you understand with? With your head? Bah!
—Nikos Kazantzakis, *Zorba the Greek*

Ninety percent of life is just being there.
—Woody Allen

1

Autonomous Agents: Walking on the Moon

1.1 Under the Volcano[1]

In the summer of 1994, an eight-legged, 1700-pound robot explorer named Dante II rapelled down a steep slope into the crater of an active volcano near Anchorage, Alaska. During the course of a six-day mission, Dante II explored the slope and the crater bed, using a mixture of autonomous (self-directed) and external control. Dante II is one product of a NASA-funded project, based at Carnegie Mellon University and elsewhere, whose ultimate goal is to develop truly autonomous robots for the purpose of collecting and transmitting detailed information concerning local environmental conditions on other planets. A much smaller, largely autonomous robot is expected to be sent to Mars in 1996, and the LunaCorp lunar rover, which is based on Dante II software, has a reserved spot on the first commercial moon shot, planned for 1997.

The problems faced by such endeavors are instructive. Robots intended to explore distant worlds cannot rely on constant communication with earth-based scientists—the time lags would soon lead to disaster. Such robots must be programmed to pursue general goals by exploring and transmitting information. For long missions, they will need to replenish their own energy supplies, perhaps by exploiting solar power. They will need to be able to function in the face of unexpected difficulties and to withstand various kinds of damage. In short, they will have to satisfy many (though by no means all) of the demands that nature made on evolving mobile organisms.

The attempt to build robust mobile robots leads, surprisingly quickly, to a radical rethinking of many of our old and comfortable ideas about the nature of adaptive intelligence.

1.2 The Robots' Parade

Elmer and Elsie

The historical forebears of today's sophisticated animal-like robots (sometimes called "animats") were a pair of cybernetic "turtles" built in 1950 by the biologist W. Grey Walter. The "turtles"—named Elmer and Elsie[2]—used simple light and touch sensors and electronic circuitry to seek light but avoid intense light. In addition, the turtles each carried indicator lights, which came on when their motors were running. Even such simple onboard equipment led to thought-provoking displays of behavior, especially when Elmer and Elsie interacted both with each other (being attracted by the indicator lights) and with the local environment (which included a few light sources which they would compete to be near, and a mirror which led to amusing, self-tracking "dancing"). In a strange way, the casual observer would find it easier to read life and purpose into the behavior of even these shallow creations than into the disembodied diagnostics of fancy traditional expert systems such as MYCIN.[3]

Herbert

One of the pioneers of recent autonomous-agent research is Rodney Brooks of the MIT Mobile Robot Laboratory. Brooks's mobile robots ("mobots") are real robots capable of functioning in messy and unpredictable real-world settings such as a crowded office. Two major characteristics of Brooks's research are the use of *horizontal* microworlds and the use of *activity-based decompositions* within each horizontal slice.

The contrast between horizontal and vertical microworlds is drawn in Clark 1989 and, in different terms, in Dennett 1978b. The idea is simple. A microworld is a restricted domain of study: we can't solve all the puzzles of intelligence all at once. A vertical microworld is one that slices off a small piece of human-level cognitive competence as an object of study. Examples include playing chess, producing the past-tense forms of English verbs, and planning a picnic, all of which have been the objects of past AI programs. The obvious worry is that when we human beings solve these advanced problems we may well be bringing to bear computational resources shaped by the other, more basic needs for which evolution equipped our ancestors. Neat, design-oriented solutions to these recent problems may thus be quite unlike the natural solutions dictated by the need to exploit existing machinery and solutions. We may be chess masters courtesy of pattern-recognition skills selected to recognize mates, food, and predators. A horizontal microworld, in contrast, is the complete behavioral competence of a whole but relatively simple creature (real or imaginary). By studying such creatures, we simplify the problems of human-level intelligence without losing sight of such biological basics as real-time response, integration of various motor and sensory functions, and the need to cope with damage.

Brooks (1991, p. 143) lays out four requirements for his artificial creatures:

A creature must cope appropriately and in a timely fashion with changes in its dynamic environment.

A creature should be robust with respect to its environment. . . .

A creature should be able to maintain multiple goals. . . .

A creature should do something in the world; it should have some purpose in being.

Brooks's "creatures" are composed of a number of distinct activity-producing subsystems or "layers." These layers do not create and pass on explicit, symbolic encodings or recodings of inputs. Instead, each layer is itself a complete route from input to action. The "communication" between distinct layers is restricted to some simple signal passing. One layer can encourage, interrupt, or override the activity of another. The resultant setup is what Brooks calls a "subsumption architecture" (because layers can subsume one another's activity but cannot *communicate* in more detailed ways).

A creature might thus be composed of three layers (Brooks 1991, p. 156):

Layer 1: Object avoidance via a ring of ultrasonic sonar sensors. These cause the mobot to *halt* if an object is dead ahead and allow reorientation in an unblocked direction.

Layer 2: If the object avoidance layer is currently inactive, an onboard device can generate random course headings so the mobot "wanders."

Layer 3: This can surpass the wander layer and instead set up a distant goal to take the mobot into a whole new locale.

A key feature of the methodology is that layers can be added incrementally, each such increment yielding a whole, functional creature. Notice that such creatures do not depend on a central reservoir of data or on a central planner or reasoner. Instead, we see a "collection of competing behaviors" orchestrated by environmental inputs. There is no clear dividing line between perception and cognition, no point at which perceptual inputs are translated into a central code to be shared by various onboard reasoning devices. This image of multiple, special-purpose problem solvers orchestrated by environmental inputs and relatively simple kinds of internal signaling is, I shall argue in a later chapter, a neuroscientifically plausible model even of more advanced brains.

Herbert,[4] built at the MIT Mobot Lab in the 1980s, exploits the kind of subsumption architecture just described. Herbert's goal was to collect empty soft-drink cans left strewn around the laboratory. This was not a trivial task; the robot had to negotiate a cluttered and changing environment, avoid knocking things over, avoid bumping into people, and identify and collect the cans. One can imagine a classical planning device trying to solve this complex real-world problem by using rich visual data to generate a detailed internal map of the present surroundings, to isolate the cans, and to plan a route. But such a solution is both costly and fragile—the environment can change rapidly (as when someone enters the room), and rich visual processing (e.g. human-level object and scene recognition) is currently beyond the reach of any programmed system.

Subsumption architectures, as we saw, take a very different approach. The goal is to have the complex, robust, real-time behavior emerge as the result of simple interactions between relatively self-contained behavior-producing subsystems. These subsystems are, in turn, controlled rather directly by properties of the encountered environment.[5] There is no central control or overall plan. Instead, the environment itself will guide the creature, courtesy of some basic behavioral responses, to success. In Herbert's case, these simple behaviors included obstacle avoidance (stopping, reorienting, etc.) and locomotion routines. These would be interrupted if a table-like outline was detected by a simple visual system. Once Herbert was beside a table, the locomotion and obstacle-avoidance routines ceded control to other subsystems that swept the table with a laser and a video camera. Once the basic outline of a can was detected, the robot would rotate until the can-like object was in the center of its field of vision. At this point, the wheels stopped and a robot arm was activated. The arm, equipped with simple touch sensors, gently explored the table surface ahead. When Herbert encountered the distinctive shape of a can, grasping behavior ensued, the can was collected, and the robot moved on.

Herbert is thus a simple "creature" that commands no stored long-term plans or models of its environment. Yet, considered as an artificial animal foraging for cans in the sustaining niche provided by the Mobot Lab ecosystem, Herbert exhibits a kind of simple adaptive intelligence in which sensors, onboard circuitry, and external environment cooperate to ensure success.

Attila

Rodney Brooks believes that robots smaller and more flexible than the lumbering Dante will better serve the needs of extraterrestrial exploration. Attila[6] weighs just $3\frac{1}{2}$ pounds and uses multiple special-purpose "mini-brains" ("finite-state machines") to control a panoply of local behaviors which together yield skilled walking: moving individual legs, detecting the forces exerted by the terrain so as to compensate for slopes, and so on. Attila also exploits infrared sensors to detect nearby objects. It is able to traverse rough terrain, and even to stand up again if it should fall on its back. Rodney Brooks claims that Attila already embodies something close to insect-level intelligence.

Periplaneta Computatrix

This is the simulated cockroach mentioned above. Beer and Chiel (1993) describe a neural-network controller for hexapod locomotion. Each leg has a mini-controller that exploits a "pacemaker" unit—an idealized model neuron whose output oscillates rhythmically. The unit will fire at intervals determined by the tonic level of excitation from a command neuron and any additional inputs it receives. The idea, borrowed from a biological model developed by K. G. Pearson (1976), is to give each leg its own rhythmic-pattern generator but then to factor in modulatory local influences involving the different sensory feedbacks from each leg as the insect traverses uneven terrain. Coordination between legs is achieved by

Figure 1.1
The first hexapod robot, built by Ken Espenschied at Case Western Reserve University under the supervision of Roger Quinn. Source: Quinn and Espenschied 1993. Reproduced by kind permission of K. Espenschied, R. Quinn, and Academic Press.

Figure 1.2
The second hexapod robot, built by Ken Espenschied at Case Western Reserve University under the supervision of Roger Quinn. Photograph courtesy of Randall Beer.

inhibitory links between neighboring pattern generators. Each leg has three motor neurons: one controls back swing, one controls forward swing, and one causes the foot to raise. The overall control circuit is again fully distributed. There is no central processor that must orchestrate a response by taking all sensory inputs into account. Instead, each leg is individually "intelligent," and simple inhibitory linkages ensure globally coherent behavior. Different gaits emerge from the interactions between different levels of tonic firing of the pacemaker units (the pattern generators) and local sensory feedback. The robot will adopt a *tripod* gait at high firing frequencies and will switch to a *metachronal* gait at lower ones. In a tripod gait, the front and back legs on one side swing in phase with the middle legs on the other side; in a metachronal gait, each leg begins its swing just after the leg behind it, in a kind of wave or ripple motion.

Although designed and tested as a pure computer simulation, the locomotion circuit has been used in a real robot body and has proved robust in the real world of friction, inertia, noise, delays, and so on. An early example of a real-world hexapod robot is shown is figure 1.1 and is further discussed in Beer and Chiel 1993 and in Quinn and Espenschied 1993. The locomotion circuit employed is also able (because it is so highly distributed) to preserved most of its functionality after damage to individual neurons or connections (Beer et al. 1992). Despite the complexity of the behavior it produces, the locomotion circuit itself is quite modest—just 37 "neurons," strategically deployed and interconnected. Nonetheless, videos of the robot hexapod and its successors provide an enthralling spectacle. One sequence shows a somewhat more complex successor robot (figure 1.2) tentatively making its way across the rough terrain provided by some fragments of polystyrene packing. A foot is extended and gently lowered. Finding no purchase (because of the local terrain), it is retracted and then placed in a slightly different location. Eventually a suitable foothold is discovered and the robot continues on its way. Such tentative exploratory behavior has all the flavor of real, biological intelligence.

Brachiation Robot

Brachiation (figure 1.3) is the branch-to-branch swinging motion that apes use to traverse highly forested terrain. Saito and Fukuda (1994) describe a robot device that learns to brachiate using a neural-network

Figure 1.3
The brachiation of a gibbon. Source: Saito and Fukuda 1994. Used by kind permission of F. Saito, T. Fukuda, and MIT Press.

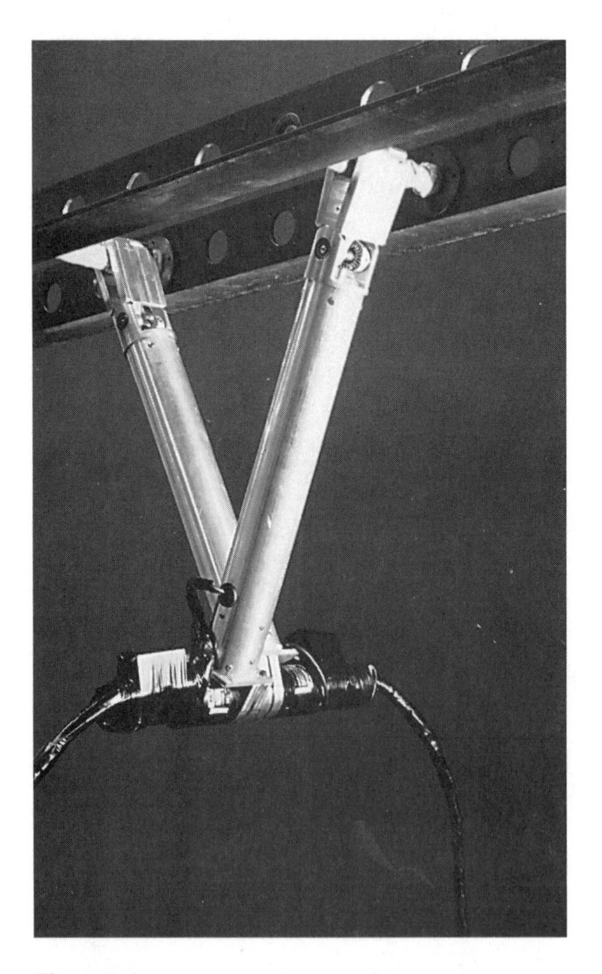

Figure 1.4
A two-link brachiation robot . Source: Saito and Fukuda 1994. Used by kind permission of F. Saito, T. Fukuda, and MIT Press.

controller. The task is especially interesting since it incorporates a learning dimension and addresses a highly time-critical behavior.

The robot uses a form of neural-network learning called *connectionist Q-learning*.[7] Q-learning involves attempting to learn the value of different actions in different situations. A Q-learning system must have a delimited set of possible actions and situations and must be provided with a reward signal informing it of the value (goodness) of a chosen action in the situation it is facing. The goal is to learn a set of situation-action pairings that will maximize success relative to a reward signal. Saito and Fukuda demonstrate that such techniques enable an artificial neural network to learn to control a two-link real-world brachiation robot (figure 1.4). The fully trained brachiation robot can swing successfully from "branch" to "branch," and if it misses it is able to use its momentum to swing back and try again.

COG

COG (Brooks 1994; Brooks and Stein 1993) must surely be the most ambitious of all the "New Robotics" projects undertaken so far. The project, spearheaded by Rodney Brooks, aims to create a high-functioning humanoid robot. The human-size robot (figure 1.5) is not mobile; it is, however, able to move its hands, arms, head, and eyes. It is bolted to a tabletop, but it can swivel at the hips. There are 24 individual motors underpinning these various degrees of freedom, and each motor has a processor devoted solely to overseeing its operation (in line with the general mobot ethos of avoiding centralized control). The arms incorporate springs, which allow some brute-mechanical smoothing. Most of the motors (excluding the eye motors) incorporate heat sensors that allow COG to gather information about its own current workings by telling it how hard various motors are working—a kind of robot version of the kinesthetic sense that tells us how our body parts are oriented in space. Each eye each comprises two cameras; one has a wide field of view with low resolution, and the other has a narrow field of view with high resolution. The cameras can move around surveying a visual scene, with the narrow-field camera mimicking the mammalian fovea. COG also receives audio information via four microphones. All this rich incoming data is processed by a "brain" composed of multiple submachines ("nodes,"

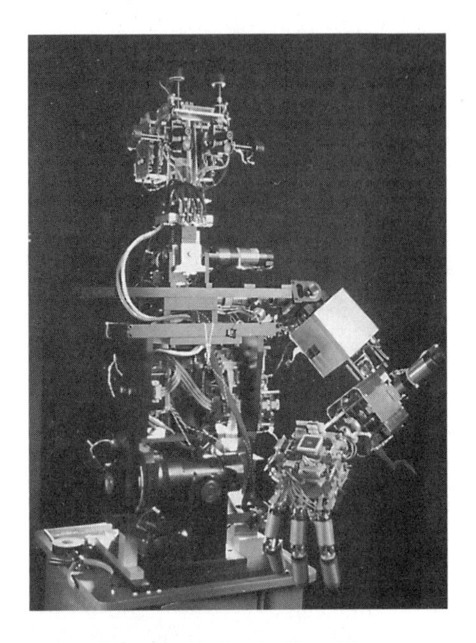

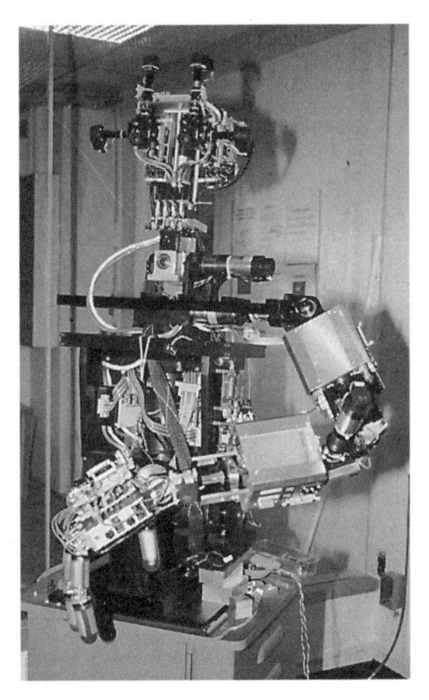

Figure 1.5
Three views of the robot COG. Photographs kindly provided by Rodney Brooks.

each with a megabyte of ROM and RAM and a dedicated operating system), which are capable of communicating with one another in some restricted ways. COG's brain is thus itself a multi-processor system, and COG's nervous system also includes other "intelligent" devices (such as the dedicated motor processors). The overall setup thus reflects much of the guiding philosophy of Brooks's work with robot insects, but it is sufficiently complex to bring new and pressing problems to the fore. Familiar features include the lack of any central memory shared by all processors, the lack of any central executive controls, the restricted communications between subdevices, and the stress on solving real-time problems involving sensing and acting. The new problems all center around the need to press coherent behaviors from such a complex system without falling back on the old, impractical methods of serial planning and central control. The ingenious strategies and tricks that enable embodied systems to maintain coherence while exploiting multiple, special-purpose, quasi-independent problem-solving routines (addressed in later chapters) shed light on the roles of language, culture, and institutions in empowering human cognition. For the moment, however, let us back off and try to extract some general morals from our parade of artificial critters.

1.3 Minds without Models

The New Robotics revolution rejects a fundamental part of the classical image of mind. It rejects the image of a *central planner* that is privy to all the information available anywhere in the system and dedicated to the discovery of possible behavioral sequences that will satisfy particular goals. The trouble with the central planner is that is profoundly impractical. It introduces what Rodney Brooks aptly termed a "representational bottleneck" blocking fast, real-time response. The reason is that the incoming sensory information must be converted into a single symbolic code so that such a planner can deal with it. And the planners' output will itself have to be converted from its propriety code into the various formats needed to control various types of motor response. These steps of translation are time-consuming and expensive.

Artificial critters like Herbert and Attila are notable for their lack of central planning. In its place the subsumption architecture puts multiple quasi-independent devices, each of which constitutes a self-contained pathway linking sensory input to action. As a result, the behaviors of such systems are not mediated by any integrated knowledge base depicting the current state of the overall environment. Such knowledge bases are often called "detailed world models," and it is a recurring theme of the new approaches that they achieve adaptive success without the use of such models.

It would be easy, however, to overstate this difference. A major danger attending any revolutionary proposal in the sciences is that too much of the "old view" may be discarded—that healthy babies may be carried away by floods of bathwater. This very danger attends, I believe, the New Roboticists' rejection of internal models, maps, and representations. Taken only as an injunction to beware the costs of central, integrated, symbolic models, the criticism is apt and important. But taken as a wholesale rejection of inner economies whose complexities include multiple action-centered representations and multiple partial world models, it would be a mistake for at least two reasons.

First, there is no doubt that the human brain does at times integrate multiple sources of information. The area that governs visual saccades (the rapid motion of the high-resolution fovea to a new target) is able to respond to multiple sensory inputs—we can saccade to the site of peripherally detected motion, to the origin of a sound, or to track an object detected only by touch. In addition, we often combine modalities, using touch, sight, and sound in complex interdependent loops where the information received in each modality helps tune and disambiguate the rest (as when we confront a familiar object in the dark corner of a cupboard).

Second, the presence of internal models intervening between input and output does not *always* constitute a time-costly bottleneck. Motor emulation provides a clean and persuasive example. Consider the task of reaching for a cup. One "solution" to a reaching problem is *ballistic* reaching. As its name implies, this style of reaching depends on a preset trajectory and does not correct for errors along the way. More skilled reaching avails itself of sensory feedback to subtly correct and guide the reaching along the way. One source of such feedback is *proprioception*, the inner sense that tells you how your body (your arm, in this case) is located in space. But proprioceptive signals must travel back from bodily peripheries to the brain, and this takes time—too much time, in fact, for the signals to be used to generate very smooth reaching movements. To solve the problem, the brain may use a trick (widely used in industrial control systems) called *motor emulation.* An emulator is a piece of onboard circuitry that replicates certain aspects of the temporal dynamics of the larger system. It takes as input a copy of a motor command and yields as output a signal identical in form to one returning from the sensory peripheries. That is, it predicts what the proprioceptive feedback should be. If the device is reliable, these predictions can be used instead of the real sensory signals so as to generate faster error-correcting activity. Such emulators are the subject of numerous detailed theoretical treatments (e.g. Kawato et al. 1987; Dean et al. 1994) that show how simple neural-network learning can yield reliable emulators and speculate on how such emulators may be realized in actual neural circuitry.

Such a motor emulator is not a bottleneck blocking real-time success. On the contrary, it facilitates real-time success by providing a kind of "virtual feedback" that outruns the feedback from the real sensory peripheries. Thus, an emulator provides for a kind of motor hyperacuity, enabling us to generate smoother and more accurate reaching trajectories than one would think possible in view of the distances and the speed of conduction governing the return of sensory signals from bodily peripheries. Yet an emulator is undoubtedly a kind of inner model. It models salient aspects of the agents' bodily dynamics, and it can even be deployed in the absence of the usual sensory inputs. But it is a partial model dedicated to a specific class of tasks. It is thus compatible with the New Roboticists' skepticism about detailed and centralized world models and with their stress on real-time behavioral success. It also underlines the intrinsic importance of the temporal aspects of biological cognition. The adaptive role of the emulator depends as much on its speed of operation (its ability to outrun the real sensory feedback) as on the information it encodes.

Carefully understood, the first moral of embodied cognition is thus to avoid excessive world modeling, and to gear such modeling as is required to the demands of real-time, behavior-producing systems.

1.4 Niche Work

The second moral follows closely from the first. It concerns the need to find very close fits between the needs and lifestyles of specific systems (be they animals, robots, or humans) and the kinds of information-bearing environmental structures to which they will respond. The idea is that we reduce the information-processing load by sensitizing the system to particular aspects of the world—aspects that have special significance because of the environmental niche the system inhabits.

We saw something of this in the case of Herbert, whose "niche" is the Coke-can-littered environment of the MIT Mobile Robot Laboratory. One fairly reliable fact about that niche is that cans tend to congregate on table tops. Another is that cans, left to their own devices, do not move or attempt to escape. In view of these facts, Herbert's computational load can be substantially reduced. First, he can use low-resolution cues to isolate tables and home in on them. Once he is at a table, he can begin a special-purpose can-seeking routine. In seeking cans, Herbert need not (and in fact cannot) form internal representations of the other objects on the table. Herbert's "world" is populated only by obstacles, table surfaces, and cans. Having located a can, Herbert uses physical motion to orient himself in a way that simplifies the reaching task. In all these respects (the use of motion, the reliance on easily detected cues, and the eschewal of centralized, detailed world models), Herbert exemplifies *niche-dependent sensing*.

The idea of niche-dependent sensing is not new. In 1934 Jakob Von Uexkull published a wonderful monograph whose title translates as *A Stroll through the Worlds of Animals and Men: A Picture Book of Invisible Worlds*. Here, with almost fairy-tale-like eloquence and clarity, Von Uexkull introduces the idea of the *Umwelt*, defined as the set of environmental features to which a given type of animal is sensitized. He describes the *Umwelt* of a tick, which is sensitive to the butyric acid found on mammalian skin. Butyric acid, when detected, induces the tick to loose its hold on a branch and to fall on the animal. Tactile contact extinguishes the olfactory response and initiates a procedure of running about until heat is detected. Detection of heat initiates boring and burrowing. It is impossible to resist quoting Von Uexkull at some length:

> The tick hangs motionless on the tip of a branch in a forest clearing. Her position gives her the chance to drop on a passing mammal. Out of the whole environment, no stimulus affects her until a mammal approaches, whose blood she needs before she can bear her young.

> And now something quite wonderful happens. Of all the influences that emanate from the mammal's body, only three become stimuli and those in definite sequence. Out of the vast world which surrounds the tick, three shine forth from the dark like beacons, and serve as guides to lead her unerringly to her goal. To accomplish this, the tick, besides her body with its receptors and effectors, has been given three receptor signs, which she can use as sign stimuli. And these perceptual cues prescribe the course of her actions so rigidly that she is only able to produce corresponding specific effector cues.
>
> The whole rich world around the tick shrinks and changes into a scanty framework consisting, in essence, of three receptor cues and three effector cues—her *Umwelt*. But the very poverty of this world guarantees the unfailing certainty of her actions, and security is more important than wealth. (ibid., pp. 11–12)

Von Uexkull's vision is thus of different animals inhabiting different *effective environments*. The effective environment is defined by the parameters that matter to an animal with a specific lifestyle. The overarching gross environment is, of course, the physical world in its full glory and intricacy.

Von Uexkull's monograph is filled with wonderful pictures of how the world might seem if it were pictured through the lens of *Umwelt*-dependent sensing (figures 1.6–1.8). The pictures are fanciful, but the insight is serious and important. Biological cognition is highly selective, and it can sensitize an organism to whatever (often simple) parameters reliably specify states of affairs that matter to the specific life form. The similarity between the operational worlds of Herbert and the tick is striking: Both rely on simple cues that are specific to their needs, and both profit by not bothering to represent other types of detail. It is a natural and challenging extension of this idea to wonder whether the humanly perceived world is similarly biased and constrained. Our third moral claims that it is, and in even more dramatic ways than daily experience suggests.

1.5 A Feel for Detail?

Many readers will surely agree that even advanced human perception is skewed toward the features of the world that matter with respect to human needs and interests. The last and most speculative of our short list of morals suggests that this skewing penetrates more deeply than we ever imagined. In particular, it suggests that our daily perceptual experiences

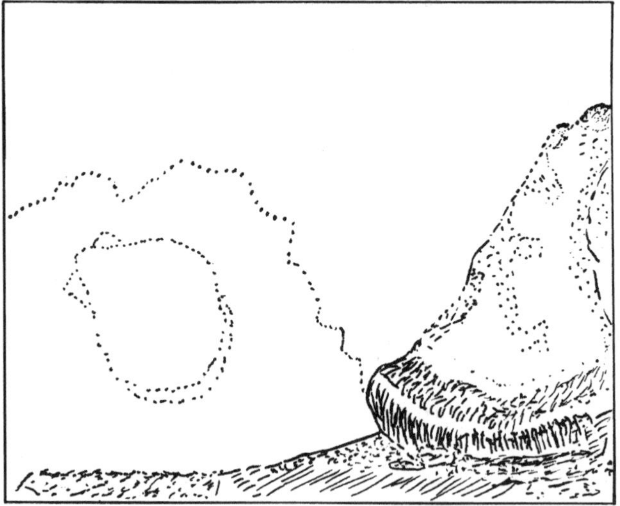

Figure 1.6
The environment and *Umwelt* of a scallop. Based on figure 19 of Von Uexkull 1934; adapted by Christine Clark, with permission of International Universities Press.

Figure 1.7
The *Umwelt* of an astronomer. Based on figure 21 of Von Uexkull 1934; adapted by Christine Clark, with permission of International Universities Press.

may mislead us by suggesting the presence of world models more durable and detailed than those our brains actually build. This somewhat paradoxical idea requires careful introduction.[8]

Consider the act of running to catch a ball. This is a skill which cricketers and baseball players routinely exhibit. How is it done? Common experience suggests that we see the ball in motion, anticipate its continuing trajectory, and run so as to be in a position to intercept it. In a sense this is correct. But the experience (the "phenomenology") can be misleading if one believes that we actively compute such trajectories. Recent research[9] suggests that a more computationally efficient strategy is to simply run so that the acceleration of the tangent of elevation of gaze from fielder to ball is kept at zero. Do this and you *will* intercept the ball before

Figure 1.8
The environment and *Umwelt* of a honeybee. Based on figure 53 of Von Uexkull 1934; adapted by Christine Clark, with permission of International Universities Press.

it hits the ground. Videotaped sequences of real-world ball interception suggest that humans do indeed—unconsciously—use this strategy. Such a strategy avoids many computational costs by isolating the minimal and most easily detectable parameters that can support the specific action of interception.

In a similar vein, an important body of research known as *animate vision* (Ballard 1991; see also P. S. Churchland et al. 1994) suggests that everyday visually guided problem solving may exploit a multitude of such tricks and special-purpose routines. Instead of seeing vision as the transformation of incoming light signals into a detailed model of a three-dimensional external world, animate-vision research investigates ways in which fast, fluent, adaptive responses can be supported by less computationally intensive routines: routines that intertwine sensing with acting and moving in the world. Examples include the use of rapid and repeated saccades to survey a visual scene and to extract detailed information only at selected foveated locations, and the exploitation of coarser cues (such as color) that can be detected at the low-resolution peripheries.

The case of rapid scanning is especially instructive. Human eyes exploit a small area (less than 0.01 percent of the overall visual field) of very high resolution. Visual saccades move this high-resolution window from point to point in a visual scene. Yarbus (1967) showed that these saccades can be intelligent in the sense that a human subject faced with an identical scene will saccade around in very different ways so as to carry out different tasks. Such saccades are very fast (about three per second) and often visit and revisit the same location. In one of Yarbus's studies, subjects were shown a picture of a room with some people in it and asked to either give the ages of the people, guess what activity they had previously been engaged in, or remember the locations of the people and objects. Very different patterns of saccade were identified, depending on which task was specified.

Frequent saccades enable us, animate-vision researchers claim, to circumvent the need to build enduring and detailed models of our visual surroundings. Instead, to borrow a slogan from Rodney Brooks, we can use the world as its own best model and visit and revisit the real-world scene, sampling it in detail at specific locations as required. The costly business of maintaining and updating a full-scale internal model of a three-dimensional scene is thus avoided. Moreover, we can sample the scene in ways suited to the particular needs of the moment.

For all that, it certainly *seems to us* as if we are usually in command of a full and detailed three-dimensional image of the world around us. But this, as several recent authors have pointed out,[10] may be a subjective illusion supported by our ability to rapidly visit any part of the scene and then retrieve detailed (but not enduring) information from the foveated region. Ballard (1991, p. 59) comments that "the visual system provides the illusion of three-dimensional stability by virtue of being able to execute fast behaviors."

A useful analogy[11] involves the sense of touch. Back in the 1960s, Mackay raised the following question: Imagine you are touching a bottle, with your eyes shut and your fingertips spread apart. You are receiving tactile input from only a few spatially separated points. Why don't you have the sensation of feeling an object with holes in it, corresponding to the spaces between your fingers? The reason is, in a sense, obvious. We use touch to *explore* surfaces, and we are accustomed to moving our fingertips so as to encounter *more* surface—especially when we know that what we are holding is a bottle. We do not treat the spaces between the sensory inputs as indicating spaces in the world, because we are used to using the senses as exploratory tools, moving first to one point and then to the next. Reflection on this case led one researcher to suggest that what we often think of as the passive sensory act of "feeling the bottle" is better understood as an action-involving cycle in which fragmentary perceptions guide further explorations, and that this action-involving cycle is the basis for the experience of perceiving a whole bottle.[12] This radical view, in which touch is cast as an exploratory tool darting hither and thither so as probe and reprobe the local environment, extends quite naturally to vision and to perception in general.

The suspicion that vision is not all it appears to be is wonderfully expressed by Patricia Churchland, V. S. Ramachandran, and Terrence Sejnowski in their 1994 paper "A critique of pure vision." In place of "picture perfect" internal representation, they too propose that we extract only a sequence of partial representations—a conjecture they characterize as the "visual semi-worlds" or "partial representations per glimpse" hypothesis. Support for such a hypothesis, they suggest, comes not only from general computational considerations concerning the use of frequent saccades and so on but also from some striking psychological experiments.[13]

The experiments involved using computer displays that "tricked" the subjects by altering the visual display during saccadic eye movements. It turned out that changes made during saccades were rather seldom detected. At these critical moments, whole objects can be moved, colors altered, and objects added, all while the subject (usually) remains blissfully unaware. Even more striking, perhaps, is related research in which a subject is asked to read text from a computer screen. The target text is never all present on the screen at once. Instead, the real text is restricted to a display of (for typical subjects) 17 or 18 characters. This text is surrounded by junk characters which do not form real words. But (and here is the trick) the window of real text moves along the screen as the subject's eyes scan from left to right. The text is nonrepetitive, as the computer program ensures that proper text systematically unfolds in place of the junk. (But, since it is a moving window, new junk appears where real text used to be.) When such a system is well calibrated to an individual subject, the subject does not notice the presence of the junk! Moreover, the subjective impression is quite distinctly one of being confronted with a full page of proper text stretching to the left and right visual peripheries. In these cases, at least, we can say with confidence that the experienced nature of the visual scene is a kind of subjective illusion caused by the use of rapid scanning and a small window of resolution and attention.

1.6 The Refined Robot

Rodney Brooks's Mobile Robot Laboratory once had the motto "Fast, cheap, and out of control." Such, indeed, is the immediate message of the New Robotics vision. Without central planning or even the use of a central symbolic code, these artificial systems fluently and robustly navigate the real world. They do so in virtue of carefully orchestrated couplings between relatively independent onboard devices and selected aspects of the environment (the robot's *Umwelt*, if you will). Despite appearances, it now seems conceivable that much of human intelligence is based on similar environment-specific tricks and strategies, and that we too may not command any central, integrated world model of the traditional style. Thus, to the extent that we take the broad morals of the New Robotics to heart, we are confronted by two immediate and pressing problems.

The first is a problem of *discovery*. If we avoid the easy image of the central planner cogitating over text-like data structures, and if we distrust our intuitions concerning what types of information we are extracting from sensory data, how should we proceed? How can we even formulate hypotheses concerning the possible structure and operation of such unintuitive and fragmentary minds? Brooks and others rely on developing a new set of intuitions—intuitions grounded in attention to specific behaviors and organized around the general idea of a subsumption architecture. As we seek to tackle increasingly complex cases, however, it is doubtful that this "handcrafting" approach can succeed. In subsequent chapters we shall investigate some ways of proceeding that seem less hostage to human intuitions: working up from real neuroscientific and developmental data, relying more on getting robot systems to learn for themselves, and even attempting to mimic genetic change so as to evolve generations of progressively more refined robots. Look to nature, and let simulated nature takes its course!

The second problem is one of *coherence*. Both the power and the puzzle of New Robotics research lie in the use of multiple, quasi-independent subsystems from which goal-directed behavior gracefully emerges under normal ecological conditions. The power lies in the robust, real-time responsiveness of such systems. The puzzle is how to maintain coherent behavior patterns as the systems grow more and more complex and are required to exhibit a wider and wider variety of behaviors. One response to such a problem is, of course, to renege on the basic vision and insist that for complex, advanced behaviors there *must* be something more like a central symbolic planning system at work. We should not, however, give up too easily. In the chapters that follow, we shall unearth a surprising number of further tricks and strategies that may induce global coherence. Most of these strategies involve the use of some type of external structure or "scaffolding" to mold and orchestrate behavior. Obvious contenders are the immediate physical environment (recall Herbert) and our ability to actively restructure that environment so as to better support and extend our natural problem-solving abilities. These strategies are especially evident in child development. Less obvious but crucially important factors include the constraining presence of public language, culture, and institutions, the inner economy of emotional response, and the various phenomena relating to group or collective intelligence. Language and culture, in particular, emerge as advanced species of external scaffolding "designed" to squeeze maximum coherence and utility from fundamentally short-sighted, special-purpose, internally fragmented minds. From its beginnings in simple robotics, our journey will thus reach out to touch—and sometimes to challenge—some of the most ingrained elements of our intellectual self-image. The Rational Deliberator turns out to be a well-camouflaged Adaptive Responder. Brain, body, world, and artifact are discovered locked together in the most complex of conspiracies. And mind and action are revealed in an intimate embrace.

2
The Situated Infant

2.1 I, Robot

Robot soda-can collectors, moon explorers, cockroaches—if all that sounded far from home, think again. The emerging perspective on embodied cognition may also offer the best hope so far for understanding central features of human thought and development. One especially promising arena is the study of infancy. The New Roboticists' vision of mind on the hoof finds a natural complement in our increasing understanding of how thought and action develop in children, for the roboticist and a growing number of developmental psychologists are united in stressing the delicate interplay of brain, body, and local environment in determining early cognitive success.

In fact (and to be historically fair), developmental psychologists were probably among the very first to notice the true intimacy of internal and external factors in determining cognitive success and change. In this respect, theorists such as Jean Piaget, James Gibson, Lev Vygotsky, and Jerome Bruner, although differing widely in their approaches, actively anticipated many of the more radical-sounding ideas now being pursued in situated robotics.[1] Nonetheless, ample scope remains for mutual illumination, since each of the two camps commands a distinct set of conceptual and experimental tools and a distinct body of data. Thus, the intellectual alliance between developmental psychology and the other sciences of the embodied mind may prove to be one of the most exciting interdisciplinary ventures of the coming decade.

This chapter explores five major landmarks along such a interdisciplinary frontier: the idea of action loops that criss-cross the organism and its environment (section 2.2), a highly interactive view of the developmental process according to which mind, body, and world act as equal partners (section 2.3); an image of biological cognition in which problem solutions often emerge without central executive control (section 2.4); recognition of the major role of external structures and support in enabling adaptive success and in pushing the envelope of individual learning (section 2.5); and an increasing skepticism, rooted in all the above considerations, concerning the ultimate value of the intuitive divisions between perception, action, and cognition (section 2.6). Cognitive development, it is concluded, cannot be usefully treated in isolation from issues concerning the child's physical embedding in, and interactions with, the world. A better image of child cognition (indeed, of *all* cognition) depicts perception, action, and thought as bound together in a variety of complex and interpenetrating ways.

2.2 Action Loops

Consider a jigsaw puzzle. One (unlikely) way to tackle such a puzzle would be to look very hard at a piece and to try to determine by reason alone whether it will fit in a certain location. Our actual practice, however, exploits a mixed strategy in which we make a rough mental determination and then physically try out the piece to see if it will fit. We do not, in general, represent the detailed shape of a piece well enough to know for certain if it is going to fit in advance of such a physical manipulation. Moreover, we may physically rotate candidate pieces even before we try to fit them, so as to simplify even the more "mentalistic" task of roughly assessing potential fit. (Recall Herbert's use of a similar procedure in which self-rotation is used to fix a can into a canonical central location in the robot's visual field.) Completing a jigsaw puzzle thus involves an intricate and iterated dance in which "pure thought" leads to actions which in turn change or simplify the problems confronting "pure thought." This is probably the simplest kind of example of the phenomena known as *action loops*.[2]

Recent developmental research by Esther Thelen and Linda Smith suggests that such interplays between thought and action may be so ubiquitous and so fundamental that these researchers suspect that all our early knowledge is built "though the time-locked interactions of perceiving and acting in particular contexts" (Thelen and Smith 1994, p. 217). To see what this means, consider the performance of infants on visual cliffs. (A visual cliff is a vertical drop covered with a strong, rigid, transparent surface, such as plexiglass.) Infants who are not yet able to crawl are demonstrably able to distinguish the shallow sides of the cliff from the area beyond the dropoff. They show increased attention and interest, and (surprisingly) they cry less on the deep side than on the shallow side. Older, more mobile infants respond to the deep side in ways associated with fear (Campos et al. 1978).[3] Clearly, both groups of infants can perceive the visual information specifying depth. The crucial difference seems to lie in how that information is put to use—how it figures in the interplay between perception and action.

Further insight into this interplay is provided by recent work on infants' responses to slopes. In this research, infants displaying different kinds of mobility (crawlers and walkers) were placed at the tops of slopes of varying degrees of steepness. The walkers (14 months) were wary of slopes of about 20° and more, and they either refused to descend or switched to a sliding mode. The crawlers dauntlessly attempted slopes of 20° and more, and usually fell as a result. (They were always caught in time.)

Under closer scrutiny, however, a fascinating pattern emerges. As crawlers increased in experience, they learned to avoid the steeper slopes. But at the point of transition, when the infants first begin to walk, this hard-won knowledge seems to have disappeared. The early walkers had to learn about steep slopes all over again. In one test, two-thirds of the new walkers "plunged without hesitation down all the steep slopes, just as they did when they first encountered them as crawlers" (Thelen and Smith 1994, p. 220).[4]

This evidence suggests not only that infants learn about the world by performing actions but also that the knowledge they acquire is itself often action-specific. Infants do not use their crawling experience to acquire knowledge about slopes in general. Rather, they acquire knowledge about how slopes figure in specific contexts involving action. Other findings concerning the context-specificity of infant knowledge point in the same general direction.[5]

This phenomenon is not restricted to infancy. Recent research on adults' mechanisms of perceptual compensation reveals a similarly action-specific profile. Thach et al. (1992) present an example involving perceptual adaptation under unusual conditions.[6] Thach and his colleagues studied human adaptation to the wearing of special glasses that shift the image to the right or the left. It is well known that the human perceptual system can learn to cope with such distortions. In fact, several experiments show that subjects can even accommodate to lenses that invert the whole visual scene so that the wearer sees an upside-down world. After wearing such glasses for a few days, subjects report sudden flips in which aspects of the world reorient themselves properly. Of course, once such adaptation has taken place, the subjects are dependent on the lenses—if they are removed, the world appears once again inverted until readaptation occurs.

What Thach's group showed was that the adaptation in the case of the sideways-shifting lenses appears to be specific to certain motor loops. Subjects were asked to throw darts at a board. At first they would miss as a result of the sideways-shifting action of the glasses. In time, however, adaptation occurred and they were able to aim as well as before. (In contrast with what happened in the experiments with lenses, this adaptation had no experiential aspect: no "backshift" in the conscious visual image was reported.) But this adaptation is, in most cases, quite motor-loop-specific. Asked to throw the darts underhand instead of overhand or to use their nondominant hand, the subjects showed no comparable improvement. Adaptation for dominant-arm, overhand throwing did not in any way carry over to the other cases. What seems to have occurred was an adaptation restricted to the specific combination of gaze angle and throwing angle used in the standard throw. What did not occur was a general, perceptual adaptation that would provide "corrected input data" for use by *any* motor or cognitive subsystem.

Thach et al. have related their results to some quite specific and fascinating hypotheses about the role of a particular neural structure—the cerebellum—in the learning of patterned responses to frequently encountered stimuli. These conjectures fit well with our emerging picture, since they suggest that the old view of the cerebellum as purely involved in motor tasks is misleading and that motor functions and some "higher" cognitive functions may be intimately bound together in the brain. For now, however, it need only be noted that some rethinking of the "passive" image of our perceptual contact with the world may be in order. In many cases, perception should not, it seems, be viewed as a process in which environmental data is passively gathered. Instead, perception may be geared, from the outset, to specific action routines.[7] The challenge, thus, is to develop "a theoretical framework that is, as it were, 'motocentric' rather than "visuocentric" (P. S. Churchland et al. 1994, p. 60). Detailed microdevelopmental studies, such as the work on slope negotiation, seem to offer a promising test bed on which to pioneer such a radical reorientation.

2.3 Development without Blueprints

A blueprint is a highly detailed plan or specification of, for example, a car or a building. The simplest (but usually the least satisfying and plausible) accounts of development depict the alteration and growth of a child's cognitive capacities as the gradual unfolding of some genetically determined "blueprint" for cognitive change. Such accounts, which dominated during the 1930s and the 1940s,[8] are neatly described by Thelen and Smith (1994, p. 6) as a vision of development as "a linear, stage-like progression through a sequence of increasingly more functional behaviors, driven towards adult forms by a grand plan (and scheduled by a grand timekeeper)." Such views are still with us, although in increasingly sophisticated forms. For example, the gradual development of walking skills is explained as an effect of maturationally determined increases in the processing speed of the brain allowing complex motor control and integration (Zelazo 1984).

From the highly interactive perspective that we have been developing, however, such approaches may be guilty of an all-too-common error. They take a complex phenomenon (e.g. the child's development of walking) and look for a single determining factor. This is what Mitchel Resnick of the MIT Media Lab calls "centralized thinking":

> . . . people tend to look for *the* cause, *the* reason, *the* driving force, *the* deciding factor. When people observe patterns and structures in the world (for example, the flocking patterns of birds or the foraging patterns of ants), they often assume centralized causes where none exist. And when people try to create patterns or structure in the world (for example, new organizations or new machines), they often impose centralized control when none is needed. (Resnick 1994, p. 120)

I have quoted this passage at length because it so perfectly captures a central message of our investigations—a message that will recur again and again in this book: Complex phenomena exhibit a great deal of self-organization. Bird flocks do not, in fact, follow a leader bird. Instead, each bird follows a few simple rules that make its behavior depend on the behavior of its nearest few neighbors. The flocking pattern emerges from the mass of these local interactions—it is not orchestrated by a leader, or by any general plan represented in the heads of individual birds. In a similar vein, certain kinds of ants forage by a process of "mass recruitment." If an ant finds food, it returns to the nest and drops a chemical marker (a pheromone) on its way. If another ant detects the trail, it will follow it to the food source. This, in turn, leads the new ant to add to the chemical trail. The stronger trail will then be more likely to attract yet another ant, which in turn finds food, adds to the chemical trail, and thus increases the trail's potency. What we thus confront is an extended process of positive feedback that soon leads to a massive concentration of activity, with hundreds of ants proceeding up and down the trail. The point is that this organization is achieved by a few simple local "rules" that, in the presence of the food source and the other ants, give rise to the apparently organized behavior.[9]

Some recent studies of infant development suggest that it, too, may be best understood in terms of the interactions of multiple local factors—factors that include, as equal partners, bodily growth, environmental factors, brain maturation, and learning. There is no "blueprint" for the behavior in the brain, or in the genes—no more than there is a blueprint for flocking in the head of the bird.

To get the flavor of the proposal, consider the case of learning to walk. The gross data are as follows: A newborn infant, when held suspended off the ground, performs well-coordinated stepping motions; at about 2 months these stepping motions are lost; the motions reappear between 8 and 10 months as the infant begins to support its weight on its feet; at about 12 months, independent walking appears. According to a "grand plan, single factor" view, we would expect these transitions to be expressions of the maturation or development of some central source—for example, the gradual capture of reflex-like processes by a higher cognitive center (see Zelazo 1984). Microdevelopmental studies suggest, how-

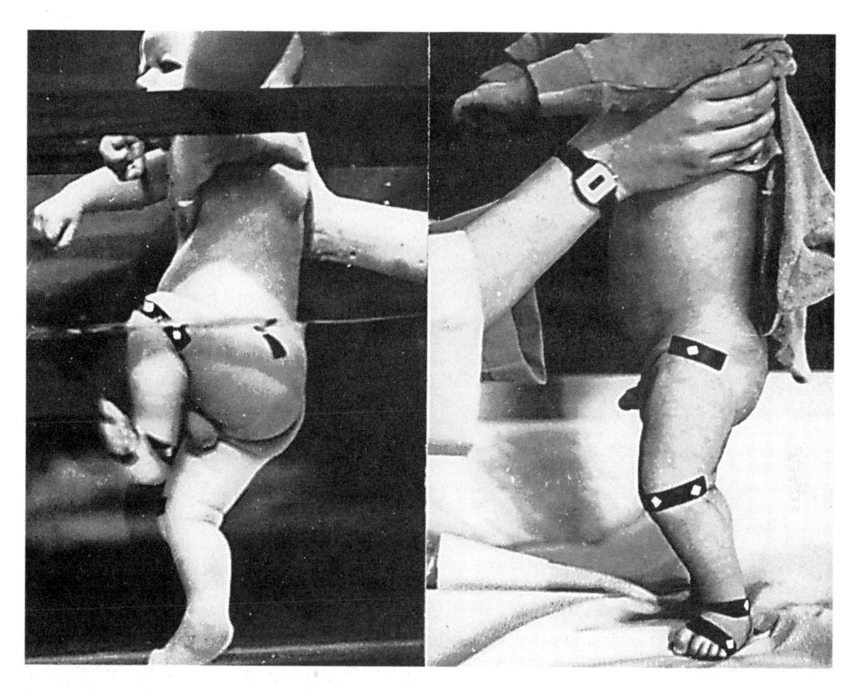

Figure 2.1
This 3-month-old infant was tested for upright stepping with his feet on the table and then when submerged in warm water. Source: Thelen and Smith 1994. Courtesy of E. Thelen, L. Smith, and MIT Press.

ever, that the transitions are not centrally orchestrated. Instead, multiple factors seem to be interacting on essentially equal terms.

For example, although reflex stepping does indeed disappear at about 2 months, nearly kinematically identical motions are still produced when the infant is lying on its back. Such "supine kicking" persists throughout the first year. The crucial parameter underlying the two-month disappearance of stepping, it now seems, is merely leg mass! In the upright position, the resistance of the leg mass at about 2 months overwhelms the spring-like action of the muscles. This hypothesis is supported by experiments (figure 2.1) in which stepping disappears after weights are added to the legs of stepping infants and by experiments in which stepping reappears after 3-month-old nonsteppers are held upright in water so that their effective leg mass is reduced.[10]

Environmental manipulations are equally effective in studying the second phase—the reappearance of stepping between 8 and 10 months. Younger, nonstepping infants, when placed on a treadmill, performed coordinated stepping; they were even able to adjust their step rate to the treadmill's speed and to adjust to asymmetric constraints when placed on a treadmill having two independent belts driven at different speeds. Treadmill stepping was found to occur in infants at all ages between 1 month and 7 months (Thelen and Smith 1994, pp. 11–17).[11]

These last results suggest a major role for a mechanical patterning caused by the backward stretching of the legs initiated by the treadmill. This component of stepping is independent of the gross normal behavioral transitions, which instead reflect the influence of multiple additional factors such as leg mass. The developmental pattern is not the expression of an inner blueprint. Rather, it reflects the complex interplay of multiple forces, some bodily (leg mass), some mechanical (leg stretching and spring-like actions), some fully external (the presence of treadmills, water, etc.), and some more cognitive and internal (the transition to volitional —i.e., deliberate—motion). To focus on any one of these parameters in isolation is to miss the true explanation of developmental change, which consists in understanding the interplay of forces in a way that eliminates the need to posit any single controlling factor.

2.4 Soft Assembly and Decentralized Solutions

A multi-factor perspective leads rather naturally to an increased respect for, and theoretical interest in, what might be termed the historical idiosyncrasies of individual development. What needs to be explained here is the delicate balance between individual variation and developmentally robust achievements. A key notion for understanding this balancing act is *soft assembly*.

A traditional robot arm, governed by a classical program, provides an example of "hard assembly." It commands a repertoire of moves, and its success depends on the precise placement, orientation, size, and other characteristics of the components it must manipulate. Human walking, in contrast, is soft-assembled in that it naturally compensates for quite major changes in the problem space. As Thelen and Smith point out, icy sidewalks, blisters, and high-heeled shoes all "recruit" different patterns of gait, muscle control, etc., while maintaining the gross goal of locomotion. Centralized control via detailed inner models or specifications seems, in general, to be inimical to such fluid, contextual adaptation. (Recall the lessons from situated robotics in chapter 1.) Multi-factor, decentralized approaches, in contrast, often yield such robust, contextual adaptation as a cost-free side effect. This is because such systems, as we saw, create actions from an "equal partners" approach in which the local environment plays a large role in selecting behaviors. In situations where a more classical, inner-model-driven solution would break down as a result of the model's incapacity to reflect some novel environment change, "equal partners" solutions often are able to cope because the environment itself helps to orchestrate the behavior.

In this vein, Pattie Maes of the MIT Media Laboratory describes a scheduling system whose goal is to match processes (jobs, or job parts) to processors (machines).[12] This is a complex task, since new jobs are always being created and since the loads of different machines continuously vary. A traditional, hard-assembled solution would invoke a centralized approach in which one system would contain a body of knowledge about the configurations of different machines, typical jobs, etc. That system would also frequently gather data from all the machines concerning their current loads, the jobs waiting, and so on. Using all this information and some rules or heuristics, the system would then search for a schedule (an efficient assignment of jobs to machines). This is the solution by Pure Centralized Cognition. Now consider, in contrast, the decentralized solution favored by Maes.[13] Here, each machine controls its own workload. If machine A creates a job, it sends out a "request for bids" to all the other machines. Other machines respond to such a request by giving estimates of the time they would require to complete the job. (A low-use machine or one that has some relevant software already loaded will outbid a heavily used or ill-prepared machines.) The originating machine then simply sends the job to the best bidder. This solution is both robust and soft-assembled. If one machine should crash, the system compensates automatically. And no single machine is crucial—scheduling is rather an emergent property of the simple interactions of posting and bidding among whatever machines are currently active. Nowhere is there a central model of the system's configuration, and hence the problems associated with updating and deploying such a model don't arise.

Soft assembly out of multiple, largely independent components yields a characteristic mix of robustness and variability. The solutions that emerge are tailored to the idiosyncrasies of context, yet they satisfy some general goal. This mix, pervasive throughout development, persists in mature problem solving and action. Individual variability should thus not be dismissed as "bad data" or "noise" that somehow obscures essential developmental patterns. Instead, it is, as Thelen and Smith insist, a powerful clue to the nature of underlying processes of soft assembly.[14]

To illustrate this, Thelen and Smith describe the development of reaching behavior in several infants. Despite the gross behavioral commonality of the final state (ability to reach), they found powerful individual differences. Reaching, in each individual case, turned out to be soft-assembled from somewhat different components, reflecting differences in the intrinsic dynamics of the infants and in their historical experience. Thelen and Smith paint a highly detailed picture; we will visit just a few highlights here.

One infant, Gabriel, was very active by nature, generating fast flapping motions with his arms. For him, the task was to convert the flapping motions into directed reaching. To do so, he needed to learn to contract muscles once the arm was in the vicinity of a target so as to dampen the flapping and allow proper contact.

Hannah, in contrast, was motorically quiescent. Such movements as she did produce exhibited low hand speeds and low torque. Her problem was not to control flapping, but to generate enough lift to overcome gravity.

Other infants present other mixtures of intrinsic dynamics, but in all cases the basic problem is one of learning to control some intrinsic dynamics (whose nature, as we have seen, can vary quite considerably) so as to achieve a goal. To do so, the central nervous system must assemble a solution that takes into account a wide variety of factors, including energy, temperament, and muscle tone. One promising proposal[15] is that in doing so the CNS is treating the overall system as something like a set of springs and masses. It is thus concerned, not with generating inner models of reaching trajectories and the like, but with learning how to modulate such factors as limb stiffness so that imparted energy will combine with intrinsic spring-like dynamics to yield an oscillation whose resting point is some desired target. That is, the CNS is treated as a control system for a body whose intrinsic dynamics play a crucial role in determining behavior.

The developmental problems that face each child are thus different, since children's intrinsic dynamics differ. What is common is the higher-level problem of harnessing these individual dynamics so as to achieve some goal, such as reaching. The job of the CNS, over developmental time, is *not* to bring the body increasingly "into line" so that it can carry out detailed internally represented commands directly specifying, e.g., arm trajectories. Rather, the job is to learn to modulate parameters (such as stiffness) which will then *interact* with intrinsic bodily and environmental constraints so as to yield desired outcomes. In sum, the task is to learn how to soft-assemble adaptive behaviors in ways that respond to local context and exploit intrinsic dynamics. Mind, body, and world thus emerge as equal partners in the construction of robust, flexible behaviors.

2.5 Scaffolded Minds

One final property of soft-assembled solutions merits explicit attention here, since it will loom large in several later chapters. It concerns the natural affinity between soft assembly and the use of *external scaffolding*. As has already been noted, the central nervous system, in learning to modulate parameters such as stiffness, was in effect solving a problem by "assuming" a specific backdrop of intrinsic bodily dynamics (the spring-like properties of muscles). Such assumed backdrops need not be confined to the agent's body. Instead, we may often solve problems by "piggy-backing" on reliable environmental properties. This exploitation of external structure is what I mean by the term *scaffolding*.

The idea of scaffolding has its roots in the work of the Soviet psychologist Lev Vygotsky.[16] Vygotsky stressed the way in which experience with external structures (including linguistic ones, such as words and sentences—see chapter 10 below) might alter and inform an individual's intrinsic modes of processing and understanding. The tradition that ensued included the notion of a zone of proximal development[17]—the idea being that adult help, provided at crucial developmental moments, would give the child experience of successful action which the child alone could not produce. Providing support for the first few faltering steps of a near-walker and supporting a baby in water to allow swimming movements would be cases in point.

The intuitive notion of scaffolding is broader, however, since it can encompass all kinds of external aid and support whether provided by adults or by the inanimate environment.[18] Two examples are the use of the physical structure of a cooking environment (grouping spices, oils, etc.) as an external memory aid (Cole et al. 1978) and the use of special eating utensils that reduce the child's freedom to spill and spear while providing a rough simulacrum of an adult eating environment (Valsiner 1987).[19] The point, for present purposes, is that environmental structures, just like the elasticity of muscles, form a backdrop relative to which the individual computational problems facing the child take shape.

Such scaffolding is common enough in noncognitive cases. The simple sponge, which feeds by filtering water, exploits the structure of its natural physical environment to reduce the amount of actual pumping it must perform: it orients itself so as to make use of ambient currents to aid its feeding.[20] The trick is an obvious one, yet not until quite recently did biologists recognize it. The reason for this is revealing: Biologists have tended to focus solely on the individual organism as the locus of adaptive structure. They have treated the organism as if it could be understood *independent of its physical world*. In this respect, biologists have resembled those cognitive scientists who have sought only inner-cause explanations of cognitive phenomena. In response to such a tendency, the biologist Vogel (1981, p. 182) has urged a principle of parsimony: "Do not develop explanations requiring expenditure of metabolic energy (e.g. the full-pumping hypothesis for the sponge) until simple physical effects (e.g. the use of ambient currents) are ruled out." The extension of Vogel's dictum to the cognitive domain is simple. It is what I once dubbed the "007 Principle":

> In general, evolved creatures will neither store nor process information in costly ways when they can use the structure of the environment and their operations upon it as a convenient stand-in for the information-processing operations concerned. That is, known only as much as you need to know to get the job done. (Clark 1989, p. 64)

This principle is reflected in the moboticists' slogan "The world is its own best representation." It is also a natural partner to ideas of soft assembly and decentralized problem solving. In place of the intellectual engine cogitating in a realm of detailed inner models, we confront the embodied, embedded agent acting as an equal partner in adaptive responses which draw on the resources of mind, body, and world. We have now seen a few preliminary examples involving bodily dynamics and the use of simple kinds of external memory store. In later chapters we shall pursue these ideas into the special realms of external structure made available by language, culture, and institutions.

2.6 Mind as Mirror vs. Mind as Controller

We have now seen a variety of ways in which cognition might exploit real-world action so as to reduce computational load. The perspective developed in the preceding sections takes us one step further, for it suggests ways in which robust, flexible behavior may depend on processes of decentralized soft assembly in which mind, body, and world act as equal partners in determining adaptive behavior. This perspective leads to a rather profound shift in how we think about mind and cognition—a shift I characterize as the transition from models of representation as mirroring or encoding to models of representation as control (Clark 1995). The idea here is that the brain should not be seen as primarily a locus of inner *descriptions* of external states of affairs; rather, it should be seen as a locus of inner *structures* that act as operators upon the world via their role in determining actions.

A lovely example of the use of such action-centered representations can be found in the work of Maja Mataric of the MIT Artificial Intelligence Laboratory. Mataric has developed a neurobiology-inspired model of how rats navigate their environments. The model has been implemented in a mobile robot. The robot rat, which has sonar sensors and a compass, achieves real-time success by exploiting the kind of subsumption architecture I described in chapter 1: it uses a set of quasi-independent "layers," each of which constitutes a complete processing route from input to output and which communicates only by passing fairly simple signals. One such layer generates boundary tracing: the robot follows walls while avoiding obstacles. A second layer detects landmarks, each of which is registered as a combination of the robot's motion and its sensory input (a corridor

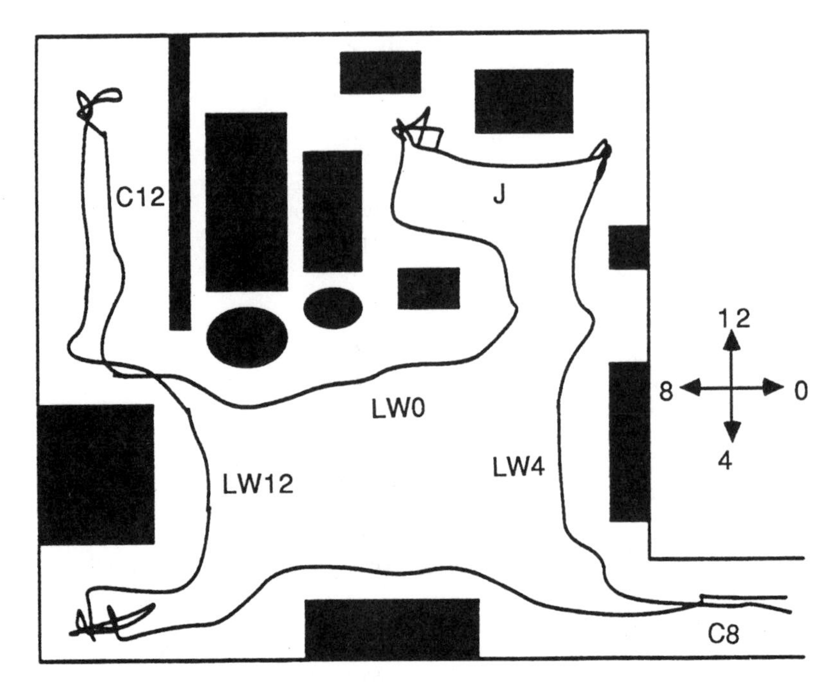

Figure 2.2
Example of a robot's reflexive navigation behavior in a cluttered office environment. Labels include landmark type and compass bearing (LW8 = left wall heading south; C0 = corridor heading north; J = long irregular boundary). Source: Mataric 1991. Used by kind permission of M. Mataric and MIT Press.

is thus remembered as the combination of forward motion and short lateral distance readings from the sonar sensors). A third layer uses this information to construct a map of the environment (figure 2.2). The map consists of a network of landmarks, each of which is, as we saw, a combination of motoric and sensory readings. All the nodes on the map process information in parallel, and they communicate by spreading activation. The robot's current location is indicated by an active node. The constructed map represents the spatial adjacency of landmarks by topological links (adjacent landmarks correspond to neighboring nodes—see figure 2.3). An active node excites its neighbors in the direction of travel, thus generating "expectations" about the next landmarks to be encountered. Suppose now that the robot wants to find its way to a remembered location. Activity at the node for that location is increased. The current

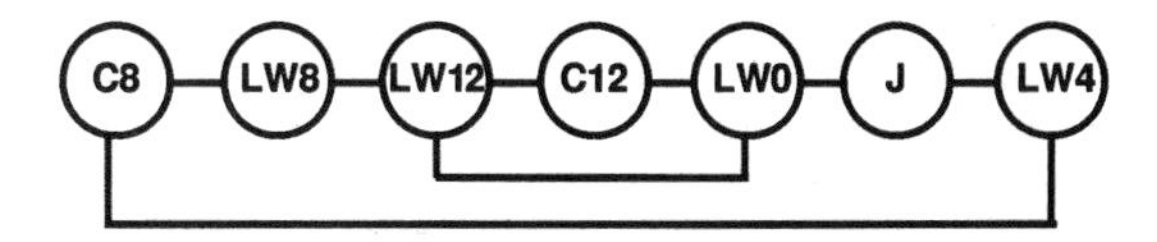

Figure 2.3
A map constructed by a robot in the environment shown in figure 2.2. Topological links between landmarks indicate physical spatial adjacency. Source: Mataric 1991. Used by kind permission of M. Mataric and MIT Press.

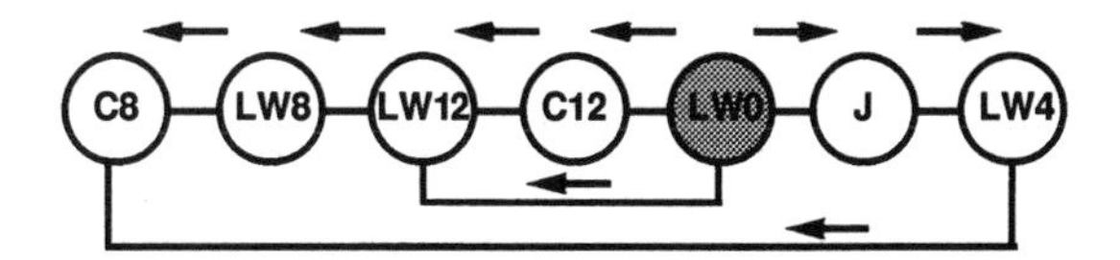

Figure 2.4
The map actively carries out path finding. Shaded node is goal node. Arrows indicate spreading of activation from goal. Source: Mataric 1991. Used by kind permission of M. Mataric and MIT Press.

location node is also active. The process of spreading activation then propagates a signal through the cognitive map, and the shortest path to the goal is computed (figure 2.4). Since the nodes on the map *themselves* combine information about the robot's movement and the correlated perceptual input, the map can *itself* act as the controller. Using the map and generating the plan for real movements turns out to be one and the same activity.

It is this feature—the ability of the map itself to act as the controller—that is of the greatest interest to us. A more classical approach would posit both some kind of stored map and a central control module that accesses the map and uses it to plan movements. The Mataric robot, in contrast, employs no reasoning device outside of the map itself. The map is its own user, and its knowledge is both descriptive (of locations) and prescriptive (it represents the relationship between two locations as the sequence of movements that would carry the robot from one landmark to the other). The robot is thus a perfect example of the idea of *action-oriented representations*: representations that simultaneously describe aspects of the world and prescribe possible actions, and are poised between pure control structures and passive representations of external reality.

A related view of internal representation was pioneered by the psychologist James Gibson (1950, 1968, 1979). This work made the mistake, however, of seeming to attack the notion of complex mediating inner states *tout court*. Despite this rhetorical slip, Gibsonian approaches are most engagingly seen only as opposing the encoding or mirroring view of internal representation.

Gibson's claim, thus sanitized, was that perception is not generally mediated by action-neutral, detailed inner-world models. It is not mediated by inner states which themselves require further inspection or computational effort (by some other inner agency) in order to yield appropriate actions. This is not, then, to deny the existence and the importance of mediating inner states altogether. Rather, it is to insist that the inner states be "action-centered"—a theme Gibson pursues by depicting organisms as keyed to detecting "affordances" in the distal environment. Such affordances are nothing other than the possibilities for use, intervention, and action offered by the local environment to a specific type of embodied agent. For example, a human perceives a chair as "affording sitting," but the affordances presented by a chair to a hamster would be radically different.

Perception, construed this way, is, from the outset, geared to tracking possibilities for action. In the place of passive re-presentation followed by inference, Gibson posits the "direct perception" of a complex of opportunities for action. In representing (as I, but not Gibson, would put it) the environment as such a complex of possibilities, we create inner states that simultaneously describe partial aspects of the world and prescribe possible actions and interventions. Such states have been aptly christened "pushmi-pullyu" representations by the philosopher Ruth Millikan.[21] Like the fabulous beast, they face both ways at once: they say how the world is *and* they prescribe a space of adaptive responses.

The common theme of these several lines of inquiry is the rejection of any blanket image of perception as the passive reception of information. Infants' perceptions of slopes, we saw, seem deeply tied to the specific motor routines by which slopes are actively engaged. Adult skill at darts appears, from the distorting-lens experiments, to involve large-scale perception/action systems rather than passive perception acting as a source of data for independent action systems to exploit. The immediate products of much of perception, such cases suggest, are not neutral descriptions of the world so much as activity-bound specifications of potential modes of action and intervention. Nor are these specifications system-neutral. Instead, as the discussion of reaching suggested, they are likely to be tailored in ways that simply assume, as unrepresented backdrop, the intrinsic bodily dynamics of specific agents. It is worth pausing to appreciate how much distance separates this vision from the classical "disembodied" image.

Perception is commonly cast as a process by which we receive information from the world. Cognition then comprises intelligent processes defined over some inner rendition of such information. Intentional action is glossed as the carrying out of commands that constitute the output of a cogitative, central system. But real-time, real-world success is no respecter of this neat tripartite division of labor. Instead, perception is itself tangled up with specific possibilities of action—so tangled up, in fact, that the job of central cognition often ceases to exist. The internal representations the mind uses to guide actions may thus be best understood as action-and-context-specific control structures rather than as passive recapitulations of external reality. The detailed, action-neutral inner models that were to provide the domain for disembodied, centralized cogitation stand revealed as slow, expensive, hard-to-maintain luxuries—top-end purchases that cost-conscious nature will generally strive to avoid.

3

Mind and World: The Plastic Frontier

3.1 The Leaky Mind

Mind is a leaky organ, forever escaping its "natural" confines and mingling shamelessly with body and with world. What kind of brain needs such external support, and how should we characterize its environmental interactions? What emerges, as we shall see, is a vision of the brain as a kind of *associative engine*, and of its environmental interactions as an iterated series of simple pattern-completing computations.

At first blush, such a vision may seem profoundly inadequate. How can it account for the sheer scale and depth of human cognitive success? Part (but only part) of the answer is that our behavior is often sculpted and sequenced by a special class of complex external structures: the linguistic and cultural artifacts that structure modern life, including maps, texts, and written plans. Understanding the complex interplay between our on-board and on-line neural resources and these external props and pivots is a major task confronting the sciences of embodied thought.

I shall begin gently, by introducing an important player to our emerging stage: the artificial neural network.

3.2 Neural Networks: An Unfinished Revolution

CYC, the electronic encyclopedia described in the introduction, was an extreme example of rule-and-symbol-style artificial intelligence. Not all projects in traditional AI were quite so gung-ho about the power of large knowledge bases and explicit encodings, but an underlying common flavor persisted throughout much of the work: the general vision of intelligence as the manipulation of symbols according to rules. "Naive Physics," for example, aimed to specify in logical form our daily knowledge about how liquids spill, how books stack, and so on (Hayes 1979). Programs like STRIPS applied theorem-proving techniques to ordinary problem solving (Fikes and Nilsson 1971), and big systems like SOAR incorporated a wide variety of such methods and representations into a single computational architecture. Nonetheless, it was not until the advent (or rebirth[1]) of so-called neural network models of mind that any *fundamentally* different proposal was put on the table.

Neural network models, as the name suggests, are at least distantly inspired by reflection on the architecture of the brain. The brain is composed of many simple processing units (neurons) linked in parallel by a large mass of wiring and junctions (axons and synapses). The individual units (neurons) are generally sensitive only to local information—each "listens" to what its neighbors are telling it. Yet out of this mass of parallel connections, simple processors, and local interactions there emerges the amazing computational and problem-solving prowess of the human brain.

In the 1980s, the field of artificial intelligence was transformed by an explosion of interest in a class of computational models that shared this coarse description of the functionality of the brain. These were the "connectionist" (or "neural network," or "parallel distributed processing") models of intelligence and cognition. The degree to which these early models resembled the brain should not be overstated.[2] The differences remained vast: the multiplicity of types of neurons and synapses was not modeled, the use of temporal properties (such as spiking frequencies) was not modeled, the connectivity was not constrained in the same ways as that of real neural systems, and so forth. Despite all this, the flavor of the models was indeed different and in a very real sense more biologically appealing. It became much easier for AI researchers working in the new paradigm to make contact with the results and hypotheses of real neuroscience. The vocabularies of the various sciences of the mind seemed at last to be moving closer together.

The basic feel of the new approach is best conveyed by example. Consider the task of pronouncing English text by turning written input (words) into phonetic output (speech). This problem can be solved by systems that encode rules of text-to-phoneme conversion and lists of exception cases, all carefully hand-coded by human programmers. DECtalk,[3] for example, is a commercial program that performs the task and whose output can drive a digital speech synthesizer. DECtalk thus relies on a fairly large, explicitly formulated, handcrafted knowledge base. NETtalk, in contrast, learns to solve the problem using an artificial neural network. The network was not provided with any set of hand-coded rules for solving the problem. Instead, it learned to solve it by exposure to a large corpus of examples of text-phoneme pairings and a learning routine (detailed below). The architecture of NETtalk was an interconnected web of units that shared some of the coarse properties of real neural networks. And the behavior of the artificial network was truly impressive. The output units were connected to a speech synthesizer, so you could hear the system slowly learning to talk, proceeding from staccato babble to half-formed words and finally to a good simulation of normal pronunciation.

NETtalk (like DECtalk) understood nothing. It was not told about the meanings of words, and it could not use language to achieve any real-world goals. But it was nonetheless a benchmark demonstration of the power of artificial neural networks to solve complex and realistic problems. How did it work?

The elements of the computational system are idealized neurons, or "units." Each unit is a simple processing device that receives input signals from other units via a network of parallel connections. Each unit sums its inputs and yields an output according to a simple mathematical function.[4] The unit is thus *activated* to whatever degree the inputs dictate, and will pass a signal to its neighbors. The signal arriving at the neighbors is determined by both the level of activation of the "sender" unit and the nature of the connection involved. Each connection has a weight, which modulates the signal. Weights can be positive (excitatory) or negative (inhibitory). The downstream signal is determined by the product of the numerical weight and the strength of the signal from the "sender" unit.

A typical connectionist network like NETtalk consists of three layers of units: "input units" (which encode the data to be processed), "hidden units" (which mediate the processing),[5] and "output units" (which specify the systems response to the data in the form of a vector of numerical activation values). The knowledge of the system is encoded in the weighted connections between the units, and it is these weights which are adapted during learning. Processing involves the spreading of activation throughout the network after the presentation of a specific set of activation values at the input units. In the case of NETtalk there are seven groups of input units, each group consisting of 29 units. Each group of 29 represents one letter, and the input consists of seven letters—one of which (the fourth) was the target whose phonemic contribution (in the context provided by the other six) was to be determined at that moment. The inputs connected to a layer of 80 hidden units, which in turn connected to 26 output units which coded for phonemes. The network involved a total of 18,829 weighted connections.

How does such a system learn? It learns by adjusting the between-unit weights according to a systematic procedure or algorithm. One such procedure is the "backpropagation algorithm." This works as follows: The system is initialized with a series of random weights (within certain numerical bounds). As they stand (being random), these weights will not support a solution of the target problem as they stand. The net is then *trained*. It is given a set of inputs, and for each input it will (courtesy of the initially random weights) produce some output —almost always incorrect. However, for each input a supervisory system sees an associated correct output (like a teacher who knows the answers in advance). The supervisory system automatically compares the actual output (a set of numerical activation values) with the correct output. For example, a face-recognition system may take as input a specification of a visual image and be required to output artificial codes corresponding to named individuals. In such a case, the correct output, for some given visual input, might be the numerical sequence ⟨1010⟩ if this has been designated as an arbitrary code for "Esther Russell." The system, courtesy of the random weights, will not do well—it may give, e.g., ⟨0. 7, 0.4, 0.2, 0.2⟩ as its initial output. At this point the supervisory system will compare the actual and desired outputs for each output unit and calculate the error on each. The errors are squared (for reasons that need not detain us) and averaged, yielding a mean squared error (MSE). The system then focuses on one weighted connection and asks whether (with all the other weights kept as they are) a slight increase or decrease in the weights would reduce the MSE. If so, then the weight is amended accordingly. This procedure is repeated for each weight; then the overall cycle of input/output/weight adjustment is repeated again and again until a low MSE is achieved. At that point, the network will be performing well (in this case, putting the right names to the visual images). Training then ceases, and the weights are frozen; the network has learned to solve the problem.[6]

This kind of learning can be usefully conceived as *gradient descent.* Imagine you are standing somewhere on the inside slopes of a giant pudding-basin-shaped crater. Your task is to find the bottom—the correct solution, the lowest error. You are blindfolded, so you cannot see where the bottom is. However, for each tiny step you might take, you can tell if the step would move you uphill (that is, in the direction of more error) or downhill (in the direction of less error). Using just this local feedback, and taking one tiny step at a time, you will inexorably move toward the bottom of the basin and then stop. Gradient-descent learning methods (of which back-propagation is an instance) proceed in essentially the same way: the system is pushed down the slope of decreasing error until it can go no further. At this point (in friendly, basin-shaped landscapes) the problem is solved, the solution reached.

Notice that at no stage in this process are the weights coded by hand. For any complex problem, it is well beyond our current capacities to find, by reflective analysis, a functional set of connection weights. What is provided is an initial architecture of so many units with a certain kind of connectivity, and a set of training cases (input-output pairs). Notice also that the upshot of learning is not, in general, a mere parrot-fashion recall of the training data. In the case of NETtalk, for example, the system learns about general features of the relation between text and spoken English. After training, the network could successfully deal with words it had never encountered before—words that were not in its initial training set.

Most important, NETtalk's knowledge of text-to-phoneme transitions does not take the form of explicit symbol-string encodings of rules or principles. The knowledge is stored in a form suitable for direct use by a brain-like system: as weights or connections between idealized "neurons" or units. The text-like forms favored by CYC and SOAR are, in contrast, forms suitable for use as external, passive knowledge structures by advanced agents such as humans. In retrospect, it is surely highly implausible that our brains (which are not so very different from those of some non-language-using creatures) should themselves use anything like the format favored by the thin projections of our thoughts onto public mediums like paper and air molecules. Brain codes must be active in a way in which text-style storage is not. The major lesson of neural network research, I believe, has been to thus expand our vision of the ways a physical system like the brain might encode and exploit information and knowledge. In this respect, the neural network revolution was surely a success.

Moreover, neural network technology looks to be with us to stay. Techniques such as those just described have been successfully applied in an incredible diversity of areas, including recognition of handwritten zip codes, visual processing, face recognition, signature recognition, robotic control, and even planning and automated theorem proving. The power and the usefulness of the technology are not to be doubted. However, its ability to illuminate biological cognition depends not just on using a processing style that is at least roughly reminiscent of real neural systems but also on deploying such resources in a biologically realistic manner. Highly artificial choices of input and output representations and poor choices of problem domains have, I believe, robbed the neural network revolution of some of its initial momentum. This worry relates directly to the emerging emphasis on real-world action and thus merits some expansion.

The worry is, in essence, that a good deal of the research on artificial neural networks leaned too heavily on a rather classical conception of the nature of the problems. Many networks were devoted to investigating what I once (Clark 1989, chapter 4; see also section 1.2 above) termed "vertical microworlds": small slices of human-level cognition, such as producing the past tense of English verbs[7] or learning simple grammars.[8] Even when the tasks looked more basic (e.g., balancing building blocks on a beam pivoting on a movable fulcrum[9]), the choice of input and output representations was often very artificial. The output of the block-balancing programs, for example, was not real motor actions involving robot arms, or even coding for such actions; it was just the relative activity of two output units interpreted so that equal activity on both indicated an expectation of a state of balance and excess activity on either unit indicated an expectation that the beam would overbalance in that direction. The inputs to the system, likewise, were artificial—an arbitrary coding for weight along one input channel and one for distance from the fulcrum along another. It is not unreasonable to suppose that this way of setting up the problem space might well lead to unrealistic, artifactual solutions. An alternative and perhaps better strategy would surely be to set up the system to take realistic inputs (e.g., from cameras) and to yield real actions as outputs (moving real blocks to a point of balance). Of course, such a setup requires the solution of many additional problems, and science must always simplify experiments when possible. The suspicion, however, is that cognitive science can no longer afford simplifications that take the real world and the acting organism out of the loop—such simplifications may obscure the solutions to ecologically realistic problems that characterize active embodied agents such as human beings. Cognitive science's aspirations to illuminate real biological cognition may not be commensurate with a continuing strategy of abstraction away from the real-world anchors of perception and action. This suspicion is, I believe, fully borne out by the significant bodies of research described in this book. One central theme which has already emerged is that abstracting away from the real-world poles of sensing and acting deprives our artificial systems of the opportunity to simplify or otherwise transform their information-processing tasks by the direct exploitation of real-world structure. Yet such exploitation may be especially essential if we hope to tackle sophisticated problem solving using the kinds of biologically plausible pattern-completing resources that artificial neural networks provide, as we shall now see.

3.3 Leaning on the Environment

Artificial neural networks of the broad stripe described above[10] present an interesting combination of strengths and weaknesses. They are able to tolerate "noisy," imperfect, or incomplete data. They are resistant to local damage. They are fast. And they excel at tasks involving the simultaneous integration of many small cues or items of information—an ability that is essential to real-time motor control and perceptual recognition. These benefits accrue because the systems are, in effect, massively parallel pattern completers. The tolerance of "noisy," incomplete, or imperfect data amounts to the ability to recreate whole patterns on the basis of partial cues. The resistance to local damage is due to the use of multiple unit-level resources to encode each pattern. The speed follows from the parallelism, as does the ability to take simultaneous account of multiple small cues.[11] Even some of the faults of such systems are psychologically suggestive. They can suffer from "crosstalk," in which similar encodings interfere with one another (much as when we learn a new phone number similar to one we already know and immediately muddle them up, thus forgetting both). And they are not intrinsically well suited to highly sequential, stepwise problem solving of the kind involved in logic and planning (Norman 1988; Clark 1989, chapter 6). A summary characterization might be "good at Frisbee, bad at logic"—a familiar profile indeed. Classical systems, with their neat, well-defined memory locations are immune to crosstalk and are excellent at logic and sequential problem solving, but they are much less well adapted to real-time control tasks.

Thus, artificial neural networks are fast but limited systems that, in effect, substitute pattern recognition for classical reasoning. As might be expected, this is both a boon and a burden. It is a boon insofar as it provides just the right resources for the tasks humans perform best and most fluently: motor control, face recognition, reading handwritten zip codes, and the like (Jordan et al. 1994; Cottrell 1991; LeCun et al. 1989). But it is a burden when we confront tasks such as sequential reasoning or long-term planning. This is not necessarily a bad thing. If our goal is to model human cognition, computational underpinnings that yield a pattern of strengths and weaknesses similar to our own are to be favored. And we *are* generally better at Frisbee than at logic. Nonetheless, we are also able, at least at times, to engage in long-term planning and to carry out sequential reasoning. If we are at root associative pattern-recognition devices,[12] how is this possible? Several factors, I believe, conspire to enable us to thus rise above our computational roots. Some of these will emerge in subsequent chapters.[13] One, however, merits immediate attention. It is the use of our old friend, external scaffolding.

Connectionist minds are ideal candidates for extensive external scaffolding. A simple example, detailed in *Parallel Distributed Processing* (the two-volume bible of neural network research[14]), concerns long multiplication. Most of us, it is argued, can learn to know at a glance the answers to simple multiplications, such as $7 \times 7 = 49$. Such knowledge could easily be supported by a basic on-board pattern-recognition device.

But longer multiplications present a different kind of problem. Asked to multiply 7222 × 9422, most of us resort to pen and paper (or a calculator). What we achieve, using pen and paper, is a reduction of the complex problem to a sequence of simpler problems beginning with 2 × 2. We use the external medium (paper) to store the results of these simple problems, and by an interrelated series of simple pattern completions coupled with external storage we finally arrive at a solution. Rumelhart et al. (1986, p. 46) comment: "This is real symbol processing and, we are beginning to think, the primary symbol processing that we are able to do. Indeed, on this view, the external environment becomes a key extension to our mind."

Some of us, of course, go on to learn to do such sums in our heads. The trick in these cases, it seems, is to learn to manipulate a mental model in the same way as we originally manipulated the real world. This kind of internal symbol manipulation is importantly distinct from the classical vision of inner symbols, for it claims nothing about the computational substrate of such imaginings. The point is simply that we can mentally simulate the external arena and hence, at times, internalize cognitive competencies that are nonetheless rooted in manipulations of the external world—cognitive science meets Soviet psychology.[15]

The combination of basic pattern-completing abilities and complex, well-structured environments may thus enable us to haul ourselves up by our own computational bootstraps. Perhaps the original vision of classical AI was really a vision of the abilities of basic pattern-completing organisms as embedded in a superbly structured environment—a vision mistakenly projected all the way back onto the basic on-board computational resources of the organism. In other words, classical rule-and-symbol-based AI may have made a fundamental error, mistaking the cognitive profile of the agent plus the environment for the cognitive profile of the naked brain (Clark 1989, p. 135; Hutchins 1995, chapter 9). The neat classical separation of data and process, of symbol structures and CPU, may have reflected nothing so much as the separation between the agent and an external scaffolding of ideas persisting on paper, in filing cabinets, or in electronic media.

The attractions of such a vision should not disguise its shortcomings. The human external environment is superbly structured in virtue of our use of linguistic, logical, and geometric formalisms and the multiple external memory systems of culture and learning. Not all animals are capable of originating such systems, and not all animals are capable of benefiting from them even once they are in place. The stress on external scaffolding thus cannot circumvent the clear fact that human brains are special. But the computational difference may be smaller and less radical than we sometimes believe. It may be that a small series of neuro-cognitive differences make possible the origination and exploitation of simple linguistic and cultural tools. From that point on, a kind of snowball effect (a positive feedback loop) may take over. Simple external props enable us to think better and hence to create more complex props and practices, which in turn "turbocharge" our thought a little more, which leads to the development of even better props. . . . It is as if our bootstraps themselves grew in length as a result of our pulling on them!

Coming back down to earth, we may pursue the idea of scaffolded pattern-completing reason in some simpler domains. Consider David Kirsh's (1995) treatment of the intelligent use of physical space. Kirsh, who works in the Cognitive Science Department at the University of California in San Diego, notes that typical AI studies of planning treat it as a very disembodied phenomenon—in particular, they ignore the way we use the real spatial properties of a work space to simplify on-board computation. Once the idea is broached, of course, examples are commonplace. Here are a few of Kirsh's favorites:

- To solve the dieter's problem of allocating $\frac{3}{4}$ of a day's allocation of cottage cheese (say, $\frac{2}{3}$ cup) to one meal, physically form the cheese into a circle, divide it into 4, and serve 3 quadrants. It is easy to see the required quantity thus arranged: not so easy to compute $\frac{3}{4}$ of $\frac{2}{3}$. (De la Rocha 1985, cited in Kirsh 1995)
- To repair an alternator, take it apart but place the pieces in a linear or grouped array, so that the task of selecting pieces for reassembly is made easier.
- To pack groceries into bags, create batches of similar items on the work surface. Grouping heavy items, fragile items, and intermediate items simplifies the visual selection process, and the relative sizes of the piles alert you to what needs accommodating most urgently.
- In assembling a jigsaw puzzle, group similar pieces together, thus allowing fine-grained visual comparison of (e.g.) all the green pieces having a straight edge.

The moral is clear: We manage our physical and spatial surroundings in ways that fundamentally alter the information-processing tasks our brains confront. (Recall the 007 Principle from chapter 2.)

What makes this cooperative approach worthwhile is the difference in nature between the kinds of computations which come naturally to the free-standing brain and the ones which can be performed by parasitizing environmental resources. But such parasitization, as we shall see, casts doubt on the traditional boundaries between mind and world themselves.

3.4 Planning and Problem Solving

There is a classical disembodied vision of planning which Phil Agre and David Chapman (1990) have labeled the "plan-as-program" idea. This is the idea (already encountered in chapter 2) of a plan as specifying a complete sequence of actions which need only be successfully performed to achieve some goal. A list of instructions for boiling an egg, or for dismantling an alternator, amounts to such a specification. A great deal of the work on "classical" planning imagines, in effect, that complex sequences of actions are determined by an internalized version of some such set of instructions. (See, e.g., Tate 1985 and Fikes and Nilsson 1971.)

Once we look closely at the real-world behaviors of planning agents, however, it becomes clear that there is a rather complex interplay between the plan and the supporting environment. This interplay goes well beyond the obvious fact that specific actions, once performed, may not have the desired effect and may thus require some on-line rethinking about how to achieve specific subgoals. In such cases the original internalized plan is still a complete, though fallible, specification of a route to success. In many cases, however, the plan turns out to be something much more partial, and much more intimately dependent on properties of the local environment.

Our earlier example of the jigsaw puzzle is a case in point. Here, an agent may exploit a strategy that incorporates physical activity in an important way. Picking up pieces, rotating them to check for potential spatial matches, and then trying them out are all *parts of the problem-solving activity*. Imagine, in contrast, a system that first solved the whole puzzle by pure thought and then used the world merely as the arena in which the already-achieved solution was to be played out. Even a system that then recognized failures of physical fit and used these as signals for replanning (a less caricatured version of classical planning) still exploits the environment only minimally relative to the rich interactions (rotations, assessments of candidate pieces, etc.) that characterize the human solution.

This crucial difference is nicely captured by David Kirsh and Paul Maglio (1994) as the distinction between *pragmatic* and *epistemic* action. Pragmatic action is action undertaken because of a need to alter the world to achieve some physical goal (e.g., one must peel potatoes before boiling them). Epistemic action, in contrast, is action whose primary purpose is to alter the nature of our own mental tasks. In such cases, we still act on the world, but the changes we impose are driven by our own computational and information-processing needs.

We have already met several examples of epistemic action, such as the use in animate vision of eye and body movements to retrieve specific types of information as and when required. What Kirsh and Maglio add to this framework is the idea that the class of epistemic actions is much broader than the animate-vision examples display. It includes all *kinds* of actions and interventions whose adaptive role is to simplify or alter the problems confronting biological brains.

A simple example, again from Kirsh (1995, p. 32), concerns the use of Scrabble tiles. During play, we physically order and reorder the tiles as a means of prompting our own on-line neural resources. Relating this to the research on artificial neural networks described in section 3.2, we may imagine the on-line neural resource as a kind of pattern-completing associative memory. One Scrabble-playing strategy is to use the special class of *external* manipulations so as to create a variety of fragmentary inputs (new letter strings) capable of prompting the recall of whole words from the pattern-completing resource. The fact that we find the external manipulations so useful suggests strongly that our on-board (in-the-head) computational resources do not themselves provide easily for such manipulations (whereas a classical AI program would find such internal operations trivial). This simple fact argues in favor of a nonclassical model of the inner resources. Once again, it looks for all the world (pun intended) as if the classical image bundles into the *machine* a set of operational capacities which in real life emerge only from the *interactions* between machine (brain) and world.

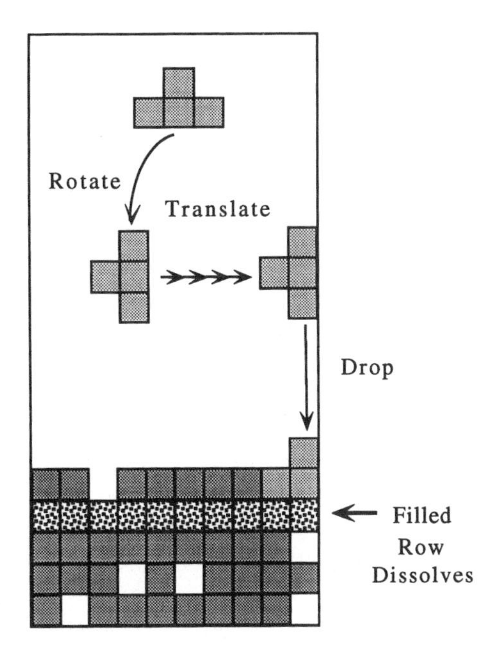

Figure 3.1
In the game Tetris, "zoids" fall one at a time from the top of the screen, eventually landing on the bottom or on zoids that have already landed. As a zoid falls, the player can rotate it, translate it to the right or the left, or immediately drop it to the bottom. When a row of squares is filled all the way across the screen, it disappears, and all rows above it drop down. Source: Kirsh and Maglio 1994. Reproduced by kind permission of D. Kirsh, P. Maglio, and Ablex Publishing Corporation.

One upshot of these observations is that external structures (including external symbols like words and letters) are special insofar as they allow types of operations not readily (if at all) performed in the inner realm.[16]

A more complex example that makes essentially the same point comes from Kirsh and Maglio's (1994) detailed studies of performance on the computer game Tetris. Tetris requires the player to place variegated geometric shapes ("zoids") into compact rows (figure 3.1). Each completed row disappears, allowing more space for new zoids. Zoids appear at the top of the screen and fall at a rate which increases as the game progresses. As a zoid falls, a player can rotate it, move it left or right, or drop it directly to the bottom. The task is thus to match shapes and geographical opportunities, and to do so subject to strong real-time constraints. One striking result of Kirsh and Maglio's investigations was that advanced players performed a variety of epistemic actions: actions whose purpose was to reduce inner computational effort rather than to gain physical ground. For example, a player may physically rotate a zoid so as to better determine its shape or to check its potential match to some geographical opportunity. Such external encounters seem faster and more reliable than internal analogues, such as imagining the zoid rotating. It is especially interesting to note (with Kirsh and Maglio) that in the case of Tetris the internal and external operations must be temporally coordinated so closely that the inner and outer systems (the brain/CNS and the on-screen operations) seem to function together as a single integrated computational unit.

The world can thus function as much more than just external memory. It can provide an arena in which special classes of external operations systematically transform the problems posed to individual brains.[17] Just as Einstein replaced the independent notions of space and time with a unified construct (space-time), Kirsh and Maglio suggest that cognitive science may need to replace the independent constructs of physical space and information-processing space with a unified physico-informational space.[18]

A final aside concerning the interplay between mind and environmental structure: Consider the case of patients with advanced Alzheimer's Disease. Many of these patients live surprisingly normal lives in the community, despite the fact that standard assessments of their capabilities suggest that many such patients should be incapable of surviving outside of special-care institutions. The key to such surprising successes, it seems, lies in the extent to which the individuals rely on highly structured environments which they create and then inhabit. These environments may incorporate multiple reminding notices around the house and strict adherence to specific routines. One patient virtually lives on a couch in the center of her apartment, since this provides a vantage point from which she can visually access the location of whatever she needs—this really *is* a case of using the world as external memory.[19]

Where does all this leave the notion of planning? The systematic problem solving of biological brains, it seems, does not really follow the plan-as-program model. Instead, individual agents deploy general strategies which incorporate operations upon the world as an intrinsic part of the problem-solving activity. Such activity can clearly involve explicitly formulated (perhaps written) plans. But even in these cases, the plan functions more like an external constraint on behavior than a complete recipe for success.[20] In a certain sense, we are like very clever mobots with Filofaxes. Our cleverness shows itself in our ability to actively structure and operate upon our environment so as to simplify our problem-solving tasks. This active structuring and exploitation extends from the simple use of spatial arrangements, through the use of specific transformations (shuffling the Scrabble tiles, rotating the Tetris zoids), all the way to the production of explicit written plans that allow easy reordering and shifting focus of attention. These latter cases involve the use of the special class of external structures that constitute maps, codes, languages, and symbols—structures that will be discussed at length in chapter 10.

3.5 After the Filing Cabinet

Artificial neural networks, we saw, provide a useful (though clearly only partial) model of some of the kinds of computational strategies that real brains seem to deploy. Such strategies stress pattern completion and associative memory at the expense of more familiar logical and symbolic manipulations. Work with artificial neural networks thus provides a valuable antidote to what has been termed the "filing cabinet" view of mind: the image of mind as a storehouse of passive language-like symbols waiting to be retrieved and manipulated by a kind of neural central processing unit. Nonetheless, some residual features of the filing-cabinet view remained unexpunged. Like a filing cabinet, mind was all too often treated as a *passive* resource: an organ for classifying and transforming incoming data but not intrinsically geared to taking action in the world. This lack of attention to the problems and possibilities attending real-world, real-time action taking manifests itself in various ways. The choice of highly abstract task domains (such as generating the past-tense forms of English verbs) and the use of very artificial forms of input and output coding are both symptoms of a vision of mind as, in essence, an organ of timeless, disembodied reason. No one thought, of course, that perception, motion, and action did not matter at all. All agreed that sooner or later such issues would have to be factored in. But it was widely believed that the additional problems such topics posed could be safely separated from the primary task of understanding mind and cognition, and that the solutions to these more "practical" problems could just be "glued onto" the computational engines of disembodied reason.

It is this methodological separation of the tasks of explaining mind and reason (on the one hand) and explaining real-world, real-time action taking (on the other) that a cognitive science of the embodied mind aims to question. Once real-world problems are confronted in their proper setting and complexity, it becomes clear that certain styles of problem solving simply will not work. And the kinds of solution that *do* work often merge the processes of reasoning and acting in unexpected ways, and cut back and forth across the traditional boundaries of mind, body, and environment.

In one sense, this should come as no surprise. Our brains evolved as controllers of bodies, moving and acting in a real (and often hostile) world. Such evolved organs will surely develop computational resources that are *complementary* to the actions and interventions they control. Thus understood, the brain need not, after all, maintain a small-scale inner replica of the world—one that supports the exact same types of operation and manipulation we customarily apply to the world. Instead, the brain's brief is to provide complementary facilities that will support the repeated exploitation of operations upon the world. Its task is to provide computational processes (such as powerful pattern completion) that the world, even as manipulated by us, does not usually afford.[21]

Where, then, is the mind? Is it indeed "in the head," or has mind now spread itself, somewhat profligately, out into the world? The question is a strange one at first sight. After all, individual brains remain the seats of consciousness and experience. But what about reason? Every thought is had by a brain. But the *flow* of thoughts and the adaptive success of reason are now seen to depend on repeated and crucial interactions with external resources. The role of such interactions, in the cases I have highlighted, is clearly computational and informational: it is to transform inputs, to simplify search, to aid recognition, to prompt associative recall, to offload memory, and so on. In a sense, then, human reasoners are truly *distributed* cognitive engines: we call on external resources to perform specific computational tasks, much as a networked computer may call on other networked computers to perform specific jobs. One implication of Kirsh and Maglio's demonstration of the role of epistemic action is thus, I suggest, a commensurate spreading of *epistemic credit*. Individual brains should not take all the credit for the flow of thoughts or the generation of reasoned responses. Brain and world collaborate in ways that are richer and more clearly driven by computational and informational needs than was previously suspected.

It would be comforting to suppose that this more integrated image of mind and world poses no threat to any of our familiar ideas about mind, cognition, and self. Comforting but false. For although specific thoughts remain tied to individual brains, the flow of reason and the informational transformations it involves seem to criss-cross brain and world. Yet it is this flow of ideas that, I suspect, we most strongly associate with the idea of the mind as the seat of reason and of the self. This flow counts for more than do the snapshots provided by single thoughts or experiences.[22] The true engine of reason, we shall see, is bounded neither by skin nor skull.

4

Collective Wisdom, Slime-Mold-Style

4.1 Slime Time

It is the spring of 1973, and the weather has been unseasonably wet. As you gaze out the window into your yard, your eye is caught by a proliferation of deep yellow blob-like masses. What could they be? Puzzled, you return to work but are unable to settle down. A while later you return to the window. The yellow jelliform masses are still in evidence, but you would swear they have moved. You are right. The newcomers are slowly but surely creeping around your yard, climbing up the nearby telephone pole—moving in on you. In a panic, you phone the police to report a likely sighting of alien life forms in the USA. In fact, what you (and many others) saw was a fully terrestrial being, but one whose life cycle is alien indeed: *Fuligo septica*, a type of acellular slime mold.[1]

Slime molds come in many varieties[2] and sizes., but all belong to the class of Mycetozoa. The name is revealing, combining 'mycet' (fungus) and 'zoa' (animal). They like moist surroundings and are often found on rotting logs, tree stumps, or piles of decaying plant matter. They are widely distributed geographically, and do not seem bound to specific climates. As one handbook puts it, "many species are apt to pop up most anywhere, unexpectedly" (Farr 1981, p. 9).

Of special interest is the life cycle of the "cellular" slime mold. Take, for instance, the species *Dictyostelium discoideum*,[3] first discovered in 1935 in North Carolina. The life cycle of *D. discoideum* begins with a so-called vegetative phase, in which the slime-mold cells exist individually, like amoeba (they are called *myxamoebae*). While local food sources last (the myxamoebae feed on bacteria) the cells grow and divide. But when

Figure 4.1
Migrating shrugs (pseudoplasmodia) of acellular slime mold. Source: Morrissey 1982. Used by permission of Academic Press.

food sources run out, a truly strange thing happens. The cells begin to cluster together to form a tissue-like mass called a *pseudoplasmodium.* The pseudoplasmodium, amazingly, is a mobile collective creature—a kind of miniature slug (figure 4.1)—that can crawl along the ground.[4] It is attracted to light, and it follows temperature and humidity gradients. These cues help it to move toward a more nourishing location. Once such a spot is found, the pseudoplasmodium changes form again, this time differentiating into a stalk and a fruiting body—a spore mass comprising about two-thirds of the cell count. When the spores are propagated, the cycle begins anew with a fresh population of myxamoebae.

How do the individual slime-mold cells (the myxamoebae) know to cluster? One solution—the biological analogue of a central planner (see chapter 3)—would be for evolution to have elected "leader cells." Such cells would be specially adapted so as to "call" the other cells, probably by chemical means, when food ran low. And they would somehow orchestrate the construction of the pseudoplasmodium. It seems, however, that nature has chosen a more democratic solution. In fact, slime-mold cells look to behave rather like the ants described in section 2.3. When food runs low, each cell releases a chemical (cyclic AMP) which attracts other cells. As cells begin to cluster, the concentrations of cyclic AMP increases, thus attracting yet more cells. A process of positive feedback thus leads to the aggregation of cells that constitutes a pseudoplasmodium. The process is, as Mitchel Resnick (1994, p. 51) notes, a nice example of what has become known as *self-organization.* A self-organizing system is one in which some kind of higher-level pattern emerges from the interactions of multiple simple components without the benefit of a leader, controller, or orchestrator.

The themes of self-organization and emergence are not, I shall suggest, restricted to primitive collectives such as the slime mold. Collectives of human agents, too, exhibit forms of emergent adaptive behavior. The biological brain, which parasitizes the external world (see chapter 3) so as to augment its problem-solving capacities, does not draw the line at inorganic extensions. Instead, the collective properties of groups of individual agents determine crucial aspects of our adaptive success.

4.2 Two Forms of Emergence

There are at least two ways in which new phenomena can emerge (without leaders or central controllers) from collective activity. The first, which I will call *direct emergence*, relies largely on the properties of (and relations between) the individual elements, with environmental conditions playing only a background role. Direct emergence can involve multiple homogeneous elements (as when temperature and pressure emerge from the interactions between the molecules of a gas), or it can involve heterogeneous ones (as when water emerges from the interactions between hydrogen and oxygen molecules). The second form of emergence, which I will call *indirect emergence*, relies on the interactions of individual elements but requires that these interactions be mediated by active and often quite complex environmental structures. The difference thus concerns the extent to which we may understand the emergence of a target phenomenon by focusing largely on the properties of the individual elements (direct emergence), versus the extent to which explaining the phenomenon requires attending to quite specific environmental details. The distinction is far from absolute, since all phenomena rely to some extent on background environmental conditions. (It can be made a little more precise by casting it in terms of the explanatory roles of different kinds of "collective variables"—see chapter 6). But we can get a working sense of the intuitive difference by looking at some simple cases.

A classic example of direct emergence is the all-too-familiar phenomenon of the traffic jam. A traffic jam can occur even when no unusual external event (such as a collision or a broken set of traffic lights) is to blame. For example, simple simulations recounted by Mitchel Resnick[5] show that bunching will occur if each car obeys just two intuitive rules: "If you see another car close ahead, slow down; if not, speed up (unless you are already moving at the speed limit)" (Resnick 1994, pp. 69, 73). Why, given just these two rules and no external obstacles, doesn't the traffic simply accelerate to the speed limit and stay there? The answer lies in the initial placements. At the start of the simulation, the cars were spaced randomly on the road. Thus, sometimes one car would start close to another. It would soon need to slow down, which would cause the car behind it to slow, and so on. The upshot was a mixture of stretches of fast-moving traffic and slow-moving jams. Every now and then a car would leave the jam, thus freeing space for the one behind it, and accelerate away. But as fast as the jam "unraveled" in one direction, it grew in the other direction as new cars reached the backmarkers and were forced to slow. Although each car was moving forward, the traffic jam itself, considered as a kind of higher-order entity, was moving backward! The higher-order structure (which Resnick calls the *collective structure*) was thus displaying behavior fundamentally different from the behavior of its components. Indeed, the individual components kept changing (as old cars left and new ones joined), but the integrity of the higher-order collective was preserved. (In a similar fashion, a human body does not comprise the same mass of matter over time—cells die and are replaced by new ones built out of energy from food. We, too, are higher-order collectives whose constituting matter is in constant flux.) Traffic jams count as cases of direct emergence because the necessary environmental backdrop (varying distances between cars) is quite minimal—random spacing is surely the default condition and requires no special environmental manipulations. The case of indirect emergence, as we shall now see, is intuitively quite different.

Consider the following scenario: You have to remember to buy a case of beer for a party. To jog your memory, you place an empty beer can on your front doormat. When next you leave the house, you trip over the can and recall your mission. You have thus used what is by now a familiar trick (recall chapter 3)—exploiting some aspect of the real world as a partial substitute for on-board memory. In effect, you have used an alteration to your environment to communicate something to yourself. This trick of using the environment to prompt actions and to communicate signals figures in many cases of what I am calling indirect emergence.

Take the nest-building behavior of some termites. A termite's building behavior involves modifying its local environment in response to the triggers provided by previous alterations to the environment—alterations made by other termites or by the same termite at an earlier time. Nest building is thus under the control of what are known as *stigmergic algorithms*.[6]

A simple example of stigmergy is the construction of arches (a basic feature of termite nests) from mudballs. Here is how it works[7]: All the termites make mud balls, which at first they deposit at random. But each ball carries a chemical trace added by the termite. Termites prefer to drop their mudballs where the chemical trace is strongest. It thus becomes likely that new mudballs will be deposited on top of old ones, which then generate an even stronger attractive force. (Yes, it's the familiar story!) Columns thus form. When two columns are fairly proximal, the drift of chemical attractants from the neighboring column influences the dropping behavior by inclining the insects to preferentially add to the side of each column that faces the other. This process continues until the tops of the columns incline together and an arch is formed. A host of other stigmergic affects eventually yield a complex structure of cells, chambers, and tunnels. At no point in this extended process is a plan of the nest represented or followed. No termite acts as a construction leader. No termite "knows" anything beyond how to respond when confronted with a specific patterning of its local environment. The termites do not talk to one another in any way, except through the environmental products of their own activity. Such environment-based coordination requires no linguistic encoding or decoding and places no load on memory, and the "signals" persist even if the originating individual goes away to do something else (Beckers et al. 1994, p. 188).

To sum up: We learn important lessons from even these simple cases of emergent collective phenomena. Such phenomena can come about in either direct or highly environmentally mediated ways. They can support complex adaptive behaviors without the need for leaders, blueprints, or central planners. And they can display characteristic features quite different in kind from those of the individuals whose activity they reflect. In the next section, we see these morals in a more familiar, human guise.

4.3 Sea and Anchor Detail

In the most successful and sustained investigation of the cognitive properties of human groups to date, Edwin Hutchins—anthropologist, cognitive scientist, and open-ocean racing sailor and navigator—has described and analyzed the role of external structures and social interactions in ship navigation. Here is his description of how some of the necessary tasks are performed and coordinated (Hutchins 1995, p. 199; my note):

> In fact, it is possible for the [navigation] team to organize its behavior in an appropriate sequence without there being a global script or plan anywhere in the system.[8] Each crew member only needs to know what to do when certain conditions are produced in the environment. An examination of the duties of members of the navigation team shows that many of the specified duties are given in the form "Do X when Y." Here are some examples from the procedures:
>
> A. Take soundings and send then to the bridge on request.
>
> B. Record the time and sounding every time a sounding is sent to the bridge.
>
> C. Take and report bearings to the objects ordered by the recorder and when ordered by the recorder.

Each member of the navigation team, it seems, need follow only a kind of stigmergic[9] procedure, waiting for a local environmental alteration (such as the placing of a specific chart on a desk, the arrival of a verbal request, or the sounding of a bell) to call forth a specific behavior. That behavior, in turn, affects the local environment of certain other crew members and calls forth further bursts of activity, and so on until the job is done.

Of course, these are *human* agents, who will form ideas and mental models of the overall process. And this general tendency, Hutchins observes, makes for a more robust and flexible system, since the individuals can monitor one another's performance (e.g., by asking for a bearing that has not been supplied on time) and, if need be (say, if someone falls ill), try to take over aspects of other jobs. Nonetheless, no crew member will have internalized all the relevant knowledge and skills.

Moreover, a large amount of work is once again done by external structures: nautical slide rules, alidades, bearing record logs, hoeys, charts, fathometers, and so on.[10] Such devices change the nature of certain computational problems so as to make them more tractable to perceptual, pattern-completing brains. The nautical slide rule, Hutchins's favorite example, turns complex mathematical operations into scale-alignment operations in physical space.[11]

Finally, and again echoing themes from chapter 3, the navigational work space itself is structured so as to reduce the complexity of problem solving. For example, the charts that will be used when entering a particular harbor are preassembled on a chart table and are laid one on top of the other in the order of their future use (the first-needed on top).

All these factors, Hutchins argues, unite to enable the overall system of artifacts, agents, natural world, and spatial organization to solve the problem of navigation. The overall (ship-level) behavior is not controlled by a detailed plan in the head of the captain. The captain may set the goals, but the sequence of information gatherings and information transformations which implement the goals need not be explicitly represented anywhere. Instead, the computational power and expertise is spread across a heterogeneous assembly of brains, bodies, artifacts, and other external structures. Thus do pattern-completing brains navigate the unfriendly and mathematically demanding seas.

4.4 The Roots of Harmony

But how does such delicate harmonization of brains, bodies, and world come about? In the cases of what I have called direct emergence the problem is less acute, for here the collective properties are determined directly by the mass action of some uniform individual propensity. Thus, if nature were (heaven forbid) to evolve cars and roads, then (given random initial distribution and the two rules rehearsed in section 4.2) traffic jams would immediately result.

Indirect emergence presents a superficially greater puzzle. In these cases, the target property (e.g., a termite nest or successful navigation of a ship) emerges out of multiple and often varied interactions between individuals and a complexly structured environment. The individuals are apparently built or designed so that the coupled dynamics of the agents and these complex environments yield adaptive success. No single individual, in such cases, needs to know an overall plan or blueprint. Yet the total system is, in a sense, well designed. It constitutes a robust and computationally economical method of achieving the target behavior. How does such design come about?

For the nervous systems of the individual termites, an important part of the answer[12] is clearly "through evolution." Hutchins suggests that a kind of quasi-evolutionary process may be at work in a navigation team too. The key feature is simply that small changes occur without prior design activity, and these changes tend to be preserved according to the degree to which they enhance biological success. Evolutionary change thus involves the gradual accretion of small "opportunistic" changes: changes which themselves alter the "fitness landscape" for subsequent changes both within the species and in other species inhabiting the same ecosystem.

Now, still following Hutchins, consider the case in which some established cognitive collective (such as a navigation team) faces a new and unexpected challenge. Suppose that this challenge calls for a fast response, so there is no time for the group to meet and reflect on how best to cope.[13] How, under such conditions, is the group to discover a new social division of labor that responds to the environmental demand? What actually happens, Hutchins shows, is that each member of the group tries to fulfill the basic functions necessary to keep the ship from going aground, but in so doing each member constrains and influences the activity of the others in what amounts to a collective, parallel search for a new yet computationally efficient division of labor. For example, one crew member realizes that a crucial addition must be performed but does not have enough time. That crew member therefore tells a nearby person to add the numbers. This in turn has effects further down the line. The solution to the problem of averting disaster emerges as a kind of equilibrium point in an iterated series of such local negotiations concerning task distribution—an equilibrium point that is determined equally by the skills of the individuals and the timing and sequence of incoming data. No crew member reflects on any overall plan for redistributing the tasks. Instead, they all do what each does best, negotiating whatever local help and procedural changes they need. In such cases there is a fast, parallel search for a coherent collective response, but the search does not involve any explicit and localized representation of the space of possible global solutions. In this sense, as Hutchins notes, the new solution is found by a process more akin to evolutionary adaptation than to global rationalistic design.

Here is a somewhat simpler version of the same idea[14]: Imagine that your task is to decide on an optimum placement of footpaths to connect a complex of already-constructed buildings (say, on a new university campus). The usual strategy is global rationalistic design, in which an individual or a small group considers the uses of the various buildings, the numbers of pedestrians, etc. and seeks some optimal pattern of linkages reflecting the patterns of likely use. An alternative solution, however, is to open the campus for business without any paths, and with grass covering all the spaces between buildings. Over a period of months, tracks will begin to emerge. These will reflect both the real needs of the users and the tendency of individuals to follow emerging trails. At the end of some period of time the most prominent trails can be paved, and the problem will have been solved without anyone's needing to consider the global problem of optimal path layout or needing to know or represent the uses of all the various buildings. The solution will have been found by means of an interacting series of small individual calculations, such as "I need to get from here to the refectory—how shall I do it?" and "I need to get to the physics lab as fast as possible—how shall I do it?" The overall effect of these multiple local decisions is to solve the global problem in a way that looks more like a kind of evolution than like classical, centralized design.

The need to account for the origins of collective success does not, it seems, force us back to the image of a central planning agency that knows the shape of the overall problem space. Instead, we may sometimes structure our own problem-solving environment as a kind of by-product of our basic problem-solving activity. On our hypothetical campus, the early walkers structure the environment as a by-product of their own actions, but subsequent walkers will then encounter a structured environment that may help them, in turn, to solve the very same problems.[15]

4.5 Modeling the Opportunistic Mind

These first few chapters have, I hope, conveyed a growing sense of the opportunistic character of much of biological cognition. For example: faced with the heavy time constraints on real-world action, and armed only with a somewhat restrictive, pattern-completing style of on-board computation, the biological brain takes all the help it can get. This help includes the use of external physical structures (both natural and artifactual), the use of language and cultural institutions (see also chapters 9 and 10 below), and the extensive use of other agents. To recognize the opportunistic and spatiotemporally extended nature of real problem solving is, however, to court a potential methodological nightmare. How are we to study and understand such complex and often non-intuitively constructed extended systems?

There is a classical cognitive scientific methodology that quite clearly *won't* do in such cases. This is the methodology of *rational reconstruction*—the practice of casting each problem immediately in terms of an abstract input-output mapping and seeking an optimal solution to the problem thus defined. Such a methodology, though perhaps never defended in principle even by workers in classical AI, nonetheless seems to have informed a large body of research.[16] Think of all those investigations of abstract microworlds: checkers, block placement, picnic planning, medical diagnosis, etc. In all such cases, the first step is to cast the problem in canonical symbolic terms and the second is to seek an efficient solution defined over a space of symbol-transforming opportunities.

Connectionists, likewise, were seen (chapter 3 above) to inherit a distressing tendency to study disembodied problem solving and to opt for abstract, symbolically defined input-output mappings.[17] Yet, from the perspectives on robotics and on infancy gained in the early chapters, it now seems more reasonable to imagine that the real-body, real-world setups of many tasks will deeply influence the nature of the problems they present to active, embodied agents. The real-world problems will be posed in a milieu that includes the spring-like properties of muscles and the presence of real, spatially manipulable objects. Such differences, as I have been at pains to show, can often make all the difference to the nature of a computational task.

In fact, the methodology of rational reconstruction can mislead in several crucial ways. First, the immediate replacement of real physical quantities with symbolic items can obscure opportunistic strategies that involve acting upon or otherwise exploiting the real world as an aid to problem solving. (Recall the 007 Principle.) Second, conceptualizing the problem in terms of an input-output mapping likewise invites a view of cognition as *passive computation*. That is, it depicts the output phase as the rehearsal of a *problem solution*. But we have now seen many cases (e.g., the strategies of animate vision and the use of the rotation button in Tetris) in which the output is an action whose role is to unearth or create further data that in turn contribute to ultimate success. These cases of what Kirsh and Maglio called "epistemic action"[18] threaten to fall through the cracks of any fundamentally disembodied, input-output vision of cognitive success. (A third threat is that the search for optimal solutions may further mislead by obscuring the role of history in constraining the space of biologically plausible solutions. Nature, as we shall see in chapter 5, is heavily bound by achieved solutions to previously encountered problems. As a result, new cognitive garments seldom are made of whole cloth; usually they comprise hastily tailored amendments to old structures and strategies.)

For all these reasons, the methodology of rational reconstruction seems to do extreme violence to the shape and nature of biological cognition. In its place, we may now glimpse the barest outlines of an alternative methodology—a methodology for studying *embodied, active cognition*. The key features of this methodology seem to be the following:

real-world, real-time focus Tasks are identified in real-world terms. Inputs are physical quantities, outputs are actions. Behavior is constrained to biologically realistic time frames.

awareness of decentralized solutions It is not simply assumed that coordinated intelligent action requires detailed central planning. Often, globally intelligent action can arise as a product of multiple, simpler interactions involving individuals, components, and/or the environment.

an extended vision of cognition and computation Computational processes are seen as (often) spread out in space and time. Such processes can extend outside the head of an individual and include transformations achieved using external props, and they can incorporate the heads and bodies of multiple individuals in collective problem-solving situations.

Thus construed, the study of embodied active cognition clearly presents some major conceptual and methodological challenges. These include (but, alas, are not exhausted by) the following:

the problem of tractability How—given this radically promiscuous view of cognition as forever leaking out into its local surroundings—are we to isolate tractable phenomena to study? Doesn't this rampant cognitive liberalism make nonsense of the hope for a genuine science of the mind?

the problem of advanced cognition How far can we really hope to go with a decentralized view of mind? Surely there is some role for central planning in advanced cognition. What, moreover, of the vision of individual reason itself? What image of rational choice and decision making is implicit in a radically emergentist and decentralized view of adaptive success?

the problem of identity Where does all this leave the individual person? If cognitive and computational processes are busily criss-crossing the boundaries of skin and skull, does that imply some correlative leakage of personal identity into local environment? Less mysteriously, does it imply that the individual brain and the individual organism are not proper objects of scientific study? These would be unpalatable conclusions indeed.

We have here a mixed bag of practical worries (How can we study the embodied embedded mind?), unsolved problems (Will the same type of story work for truly advanced cognition?), and conceptual anomalies (Does leaky cognition imply leaky persons? Are brains somehow improper objects of study?). In the remaining chapters, I shall address all these issues. In particular, I shall try to respond in detail to the methodological and practical worries (chapters 5–7), to clarify the conceptual problems (chapters 6 and 8), and to begin to address the pressing problem of advanced cognition (chapters 9 and 10). The key to integrating the facts about advanced cognition with the vision of embodied active cognition lies, I shall suggest, in better understanding the roles of two very special external props or scaffolds: language and culture.

In sum: The death of rational reconstruction creates something of a conceptual and methodological vacuum. Our remaining task is to fill the void.

Intermission: A Capsule History

Cognitive science, as sketched in the preceding chapters, can be seen in terms of a three-stage progression. The first stage (the heyday of classical cognitivism) depicted the mind in terms of a central logic engine, symbolic databases, and some peripheral "sensory" modules. Key characteristics of this vision included these ideas:

memory as retrieval from a stored symbolic database,
problem solving as logical inference,
cognition as centralized,
the environment as (just) a problem domain,

and

the body as input device.

The connectionist (artificial neural network) revolution took aim at the first three of these characteristics, replacing them with the following:

memory as pattern re-creation,
problem solving as pattern completion and pattern transformation,

and

cognition as increasingly decentralized.

This radical rethinking of the nature of the inner cognitive engine, however, was largely accompanied by a tacit acceptance of the classical marginalization of body and world. It is this residual classicism which the kind of research reported earlier confronts head on. In this research, the most general tenets of the connectionist view are maintained, but they are augmented by a vision of

the environment as an active resource whose intrinsic dynamics can play important problem-solving roles and

the body as part of the computational loop.

To thus take body and world seriously is to invite an *emergentist* perspective on many key phenomena—to see adaptive success as inhering as much in the complex *interactions* among body, world, and brain as in the inner processes bounded by skin and skull. The challenges for such an approach, however, are many and deep. Most crucial is the pressing need to somehow balance the treatment of the internal (brain-centered) contribution and the treatment of external factors in a way that does justice to each. This problem manifests itself as a series of rather abstract-sounding worries—but they are worries with major concrete consequences for the conduct and the methodology of a science of the embodied mind. These worries include

finding the right vocabulary to describe and analyze processes that criss-cross the agent/environment boundary,

isolating appropriate large scale systems to study and motivating some decomposition of such systems into interacting component parts and processes,

and

understanding familiar terms such as 'representation', 'computation', and 'mind' in ways which fit the new picture (or else rejecting such terms entirely).

In short: How should we *think* about the kind of phenomena we have displayed—and how many of our old ideas and prejudices will we have to give up to do so? This is the topic of part II.

II

Explaining the Extended Mind

Our own body is in the world as the heart is in the organism . . . it forms with it a system.

—Maurice Merleau-Ponty, *Phenomenology of Perception*; passage translated by David Hilditch in his Ph.D. thesis, At the Heart of the World (Washington University, 1995)

5
Evolving Robots

5.1 The Slippery Strategems of the Embodied, Embedded Mind

How should we study the embodied, embedded mind? The problem becomes acute once we realize that nature's solutions will often confound our guiding images and flout the neat demarcations (of body, brain, and world) that structure our thinking. The biological brain is, it seems, both constrained and empowered in important and sometimes non-intuitive ways. It is constrained by the nature of the evolutionary process—a process that must build new solutions and adaptive strategies on the basis of existing hardware and cognitive resources. And it is empowered, as we have seen, by the availability of a real-world arena that allows us to exploit other agents, to actively seek useful inputs, to transform our computational tasks, and to offload acquired knowledge into the world.

This combination of constraints and opportunities poses a real problem for the cognitive scientist. How can we model and understand systems whose parameters of design and operation look (from an ahistorical, disembodied design perspective) so messy and non-intuitive? One partial solution is to directly confront the problem of real-world, real-time action, as in the robotics work surveyed in chapter 1. Another is to attend closely to the interplay between cognition and action in early learning, as in the developmental research discussed in chapter 2. An important additional tool—the focus of the present chapter—is the use of *simulated evolution* as a means of generating control systems for (real or simulated) robots. Simulated evolution (like neural network learning) promises to help reduce the role of our rationalistic prejudices and predispositions in the search for efficient solutions.

5.2 An Evolutionary Backdrop

Naturally evolved systems, it has often been remarked, simply do not function the way a human designer might expect.[1] There are several reasons for this. One, which we have already seen exemplified many times over, involves a propensity for distributed solutions. The now-familiar point is that where a human designer will usually build any required functionality directly into a distinct device for solving a given problem, evolution is in no way constrained by the boundaries between an organism or device and the environment. Problem solving easily becomes distributed between organism and world, or between groups of organisms. Evolution, having in a very real sense no perspective on a problem at all, is not prevented from finding cheap, distributed solutions by the kinds of blinkers (e.g., the firm division between device and operating domain) that help human engineers focus their attention and decompose complex problems into parts.

This is not, however, to suggest that principles of decomposition play no role in natural design. But the kind of decomposition that characterizes design by natural selection is a very different beast indeed. It is a decomposition dictated by the constraint of *evolutionary holism*—a principle, explicitly formulated by Simon (1969), which states that complex wholes will usually be developed incrementally over evolutionary time, and that the various intermediate forms must themselves be whole, robust systems capable of survival and reproduction. As Dawkins (1986, p. 94) puts it, the key is to think in terms of *trajectories* or *paths* though evolutionary time, with whole successful organisms as steps along the way.

This is a strong constraint. A wonderfully adaptive complex design that lacks any such evolutionary decomposition (into simpler but successful ancestral forms) will never evolve. Moreover, the transitions between forms should not be too extreme: they should consist of small structural alterations, each of which yields a whole, successful organism.

One story has it, for example,[2] that our lungs evolved from a foundation provided by the swim bladders of fish. Swim bladders are sacs of air that facilitate movement in watery environments. It has been suggested that our current susceptibility to pleurisy and emphysema can be traced to features of the swim-bladder adaptation. Lieberman (1984, p. 22) is thus led to comment that "swim bladders are logically designed devices for swimming—they constitute a Rube Goldberg system for breathing."

The moral is an important one. It is that the constraints of evolutionary holism, coupled with the need to proceed via small incremental changes to existing structures, can yield solutions to current problems that owe a great deal to their particular historical antecedents. As the cell geneticist François Jacob (1977, p. 1163) put it: "Simple objects are more dependent on (physical) constraints than on history. As complexity increases, history plays the greater part." Jacob likens evolution to a tinkerer rather than an engineer. An engineer sits down at a blank drawing board and designs a solution to a new problem from scratch; a tinkerer takes an existing device and tries to adapt it to some new purpose. What the tinkerer produces may at first make little sense to the engineer, whose thinking is not constrained by available devices and ready-to-hand resources. Natural solutions to the problems faced by complex evolved creatures may likewise appear opaque from a pure, ahistorical design perspective.

One way to begin to understand such initially opaque, historically path-dependent and opportunistic problem solutions is to try artificially to recapitulate the evolutionary process itself: set a tinkerer to catch a tinkerer. Enter the genetic algorithm.

5.3 Genetic Algorithms as Exploratory Tools

Biological evolution, as we all know, works by a process of diversification and selection. Given some population of organisms, and given variety within that population, some will do better at survival and reproduction than others. Add to this a mechanism of transmission, which causes the descendants of the fittest to inherit some of the structure of their forebears, and the minimal conditions for evolutionary search are in place. Transmission normally involves inbuilt means of further variation (e.g., mutation) and diversification (e.g., the splitting and recombination processes characteristic of sexual reproduction). By an iterated sequence of variations, diversifications, selections, and transmissions, the evolutionary process performs a search in the space of structural options—a search that will tend to zero in on the fitter solutions to the problems of survival and reproduction.

Genetic algorithms[3] simulate this kind of evolutionary process. The population initially consists of a variety of software individuals, either hand coded or randomly generated. Such "individuals" might be lines of code, data structures, whole hierarchical computer programs, neural networks, or whatever. The individuals are then allowed to behave—to act in some environment in ways that will allow the computation, after some time, of a measure of fitness for each one. (How much food did it find? Did it avoid predators? . . .) The initial coding for the fittest individuals (usually stored as binary strings) is then used as a basis for reproduction (i.e., for generating the next population). But instead of simply copying the most successful individuals, operations of crossover and mutation are employed. In mutation, a small random change is made to the structure of the coding for an individual. For example, if the individual is a neural network, a few weights might be subtly varied. In crossover, parts of the codings for two individuals are recombined so as to mimic the rough dynamics of sexual reproduction. The new generation is thus based on the most successful variants among the old, but continues the process of searching for efficient solutions by investigating some of the space surrounding the previous good solutions. This process, when iterated over hundreds of thousands of generations, constitutes (in certain problem domains) a powerful version of gradient-descent search[4]—except that here the learning increments occur generation by generation instead of during an individual lifetime.

Such techniques have been used to evolve problem solutions in a wide variety of domains, from trail following in artificial ants (Jefferson et al. 1990; Koza 1991), to discovering laws of planetary motion (Koza 1992), to evolving neural network controllers for artificial insects (Beer and Gallagher 1992). The latter kind of use is especially interesting insofar as it allows us to study the effects of incremental evolutionary learning in settings that include rich bodily and environmental dynamics, as we shall now see.

5.4 Evolving Embodied Intelligence

Walking, seeing, and navigating are fundamental adaptive strategies exploited by many evolved creatures. Can simulated evolution help us to understand them better? The answer looks to be a tentative Yes.

Consider walking. Randall Beer and John Gallagher (1992) have used genetic algorithms to evolve neural network controllers for insect locomotion. These evolved controllers turn out to exploit a variety of robust and sometimes nonobvious strategies. Many of these strategies rely on close and continued interactions between the controller and the environment and do not involve the advance construction of detailed and explicit motor programs. Moreover, the best of the controllers were able to cope with a variety of challenging situations, including operation with and without sensory feedback and including automatic compensation for certain kinds of structural change.

Beer and Gallagher's robot insect was a kind of simulated cockroach[5] with six legs. Each leg was jointed and could have its foot up or down. A sensor on each leg reported the angle of the legs relative to the body. The simulated insect was controlled by a network of neural nets (each leg had a dedicated five-neuron network controller). Each five-neuron subnet included three motor neurons driving the leg and two "extra" neurons whose role was left open. Each subnet received input from the sensor associated with the leg it controlled. A genetic algorithm (see section 5.3) was used to discover a set of features (such as connection weights—see chapter 3) that would enable this kind of control architecture to generate stable and robust locomotion. This, in turn, involved finding weights, biases, and time constants (response speeds) capable of generating a viable motion pattern for each leg, and also coordinating the motions of all the legs.

Beer and Gallagher evolved eleven controllers, each of which used a different set of weights and parameter values. All the controllers produced good locomotion, and all used the "tripod gait" favored by real fast-walking insects.

The importance of the controller-environment interaction was demonstrated by evolving solutions in three different settings. In the first setting, evolutionary search occurred with leg sensors operative. Under these conditions, unsurprisingly, the final solutions relied heavily on continued sensory feedback. If the sensor were subsequently disabled, locomotion was lost or badly disrupted. In the second setting, evolutionary search occurred without sensory feedback. Under these "blind" conditions, solutions were discovered that relied only on central pattern generators and hence produced a somewhat clumsy but reliable locomotion akin to that of a toy robot.

More interesting by far were the results obtained when sensory feedback was *intermittently* present during evolutionary search. Under these uncertain conditions, controllers evolved that could produce smooth walking using sensory feedback when available, switch to "blind" pattern generation in the absence of sensory feedback (and hence produce viable albeit less elegant locomotion), and even compensate automatically for certain structural changes (e.g., alterations of leg length, such as occur during biological growth). The explanation of this last property involves the modulation exercised by sensory feedback on the pattern generator in these "mixed" solutions. The altered leg length affects the readings at the sensor, and this causes a commensurate slowing of the motor output generator. This kind of automatic compensation has a biologically realistic flavor—think of how a cat automatically adopts a new three-legged gait when one leg is injured, or how a human being adapts to walking on an icy surface or with a sprained ankle. Yet, as Beer (1995b) points out, this kind of adaptation is not a result of individual learning as such—rather, the adaptation is inherent in the original dynamics of the system, and the new situation (damage, leg growth, or whatever) merely causes it to be displayed.

Overall, the kind of solution embodied in the mixed controller involves such a subtle balancing of central pattern generation and sensory modulation that, Beer suggests, the design might easily have eluded a human analyst. By using the genetic algorithm, solutions can be found that truly make the most of whatever environmental structure is available and which are not hobbled by our natural tendency to seek neat, clean, easily decomposable problem solutions. Of course, the bad news about messier, more biologically realistic and interactive solutions is that they are not just hard to discover but also hard to *understand* once we have them. We shall return to this problem in section 5.7.

Further experiments echo Beer and Gallagher's results in other domains. Harvey et al. (1994) evolved control systems for visually guided robots, and Yamuchi and Beer (1994) have evolved networks capable of controlling a robot that used sonar input to perform landmark recognition and navigation. Johnson et al. (1994) used genetic programming to evolve animate-vision-style routines for the computationally cheap solution of ecologically realistic visual processing tasks (recall chapter 1), finding evolved solutions which significantly outperformed the best programs they were able to produce by hand. There is thus ample evidence of the power of simulated evolutionary search to unearth robust and unobvious solutions to biologically realistic problems. This optimistic statement, however, should be tempered by the recognition of several severe limitations that afflict most of the work in this field. The most important of those limitations are the "freezing" of the problem space, the use of fixed neural and bodily architectures, the lack of a rich phenotype/genotype distinction, and the problem of "scaling up" in evolutionary search.

By the "freezing" of the problem space I mean the tendency to predetermine a fixed fitness function and to use simulated evolution merely to maximize fitness relative to this preset goal (walking, navigating, or whatever). This approach ignores one of the factors that most strongly differentiate real evolutionary adaptation from other forms of learning: the ability to *coevolve* problems and solutions. A classic example is the coevolution of pursuit and evasion techniques in animal species.[6] The crucial point is just that natural evolution does not operate so as to "solve" a fixed problem. Instead, the problems themselves alter and evolve in a complex web of coevolutionary change.

Equally problematic is the tendency to search a problem space partially defined by some fixed bodily or neural architecture. Once again, these searches freeze parameters that, in the natural world, are themselves subject to evolutionary change. For example, the simulated cockroach had a fixed bodily shape and a fixed set of neural resources. Real evolutionary search, in contrast, is able to vary both bodily shape[7] and gross neural architecture.

Another biological distortion involves the use of rather direct genotype-phenotype mappings. In standard genetic-algorithm search, the new populations of individuals are fully specified by their genotypes. In contrast, the way real genes become expressed in real bodies allows a much greater role for environmental interactions over individual developmental time. In fact, the image of genes "coding for" physical features is often quite misleading. Rather, genes code for possible physical features, in ways that depend heavily on a variety of environmental factors which affect their expression. The capacity to select genetic factors whose ultimate expression in individuals remains under a large degree of environmental control allows biological evolution to exploit several degrees of freedom not present in most artificial models.[8]

Finally, there is a widely acknowledged problem of "scaling up." Most of the work reported above uses genetic search applied to relatively small neural network controllers. As the number of parameters characterizing the controllers increases, standard varieties of evolutionary search become increasingly inefficient. The key to overcoming this problem seems to lie in some combination of better genetic encodings and the "offloading" of some of the burden onto the environment (i.e., reducing the amount of information encoded in the genotype by relying on developmental interactions with a structuring environment). In this way, the scaling problem and the previous phenotype/genotype problem may be more closely linked than is initially apparent.[9]

Clearly, then, the use of simulated evolution is far from being a panacea for autonomous-agent research. Nonetheless, such methods have already won a place in the tool kit of the cognitive sciences of the embodied mind. Exactly how central a place will depend also on the resolution of a rather vigorous in-house dispute concerning the legitimacy and the value of using *simulated* agents and environments in understanding embodied, active cognition.

5.5 SIM Wars (Get Real!)

Artificial evolution takes place, by and large, in populations of simulated organisms attempting to negotiate simulated environments. But the use of simulations is itself a point of contention within the community of researchers studying embodied, embedded cognition. On the one hand, the use of simulated worlds and agents provides clear benefits in terms of problem simplification and the tractability of studying large populations. On the other hand, one of the major insights driving much autonomous-agent research is precisely a recognition of the unsuspected complexity of real agent-environment interactions and of the surprising ways in which real-world features and properties can be exploited by embodied beings. Fans of real-world robotics[10] note that researchers routinely underestimate the difficulty of problems (by ignoring such real-world features as noise and the unreliability of mechanical parts) and also fail to spot quick and dirty solutions that depend on such gross physical properties as the elasticity and "give" of certain parts.[11]

A useful nonrobotic example of the role of such physical properties is developed by Tim Smithers (1994, pp. 64–66) in his account of "hunting," a phenomenon associated with the second generation of fly-ball governors used to regulate the power outputs of early steam engines. Fly-ball governors (also known as Watt governors after their inventor, James Watt) are used to maintain constant speed in a flywheel, run off a steam engine, to which other machinery is connected. Without governance, the speed of the flywheel varies according to steam fluctuations, workload alterations, and other factors. The governor is based on a vertical spindle geared to the main flywheel. The spindle has two arms, attached by hinges, each of which has a metal ball at the end. The arms swing out as the flywheel turns, to a degree determined by the speed of rotation. The arms directly operate a throttle valve that reduces the flow of steam as the arms raise (and hence as the speed of the flywheel increases) and increases it as the arms lower (and hence as the speed of the flywheel decreases). This arrangement maintains a constant speed of rotation of the flywheel, as is required for many industrial applications. With increased precision of manufacture, Smithers notes, a new generation of governors began to exhibit a problem not seen in the earlier, "cruder" versions. The new, finely machined governors would often fail to determine a single fixed speed of rotation, and would instead oscillate between slowing down and speeding up. This "hunting" for a constant speed occurred because the new governors were reacting too quickly to the main shaft's speed and thus, in effect, overcorrecting each time. Why did the early, crude versions outperform their finely engineered successors? The reason was that friction between joints, bearings, and pulleys was, in the early versions, sufficient to damp the system's responses, thus protecting it from the looping cycles of rapid overcompensation observed in the newer machines. Modern regulators, we are told, rely on additional components to prevent hunting, but these pay a price in being more difficult to set up and use (ibid., p. 66).

Smithers shows that attempts to fine tune the sensory systems of simple real-world robots can run into similar problems. If robot behavior depends closely on sensor readings, highly sensitive devices can become overresponsive to small perturbations caused by relatively insignificant environmental changes, or even by the operation of the sensor itself. Increased resolution is thus not always a good thing. By using less accurate components, it is possible to design robots in which properties of the physical device (e.g., mechanical and electrical losses) act so as to damp down responses and hence avoid undesirable variations and fluctuations. As a result, Smithers suggests, it may even be misleading to think of the sensors as measuring devices—rather, we should see them as filters whose role is, in part, to soak up behaviorally insignificant variations so as to yield systems able to maintain simple and robust interactions with their environment. Real physical components, Smithers argues, often provide much of this filtering or sponge-like capacity "for free" as a result of mechanical and electrical losses inherent in the physical media. These effects, clearly, will not be available "for free" in simulated agent-environment systems. Simulation-based work is thus in danger of missing cheap solutions to important problems by failing to recognize the stabilizing role of gross physical properties such as friction and electrical and mechanical loss.

Another problem with a pure simulation-based approach is the strong tendency to oversimplify the simulated environment and to concentrate on the intelligence of the simulated agent. This furthers the deeply misguided vision of the environment as little more than the stage that sets up a certain problem. In contrast, the arguments of the previous chapters all depict the environment as a rich and active resource—a partner in the production of adaptive behavior. Related worries include the relative poverty of the simulated physics (which usually fails to include crucial real-world parameters, such as friction and weight), the hallucination of perfect information flow between "world" and sensors, and the hallucination of perfectly engineered and uniform components[12] (e.g., the use of identical bodies for all individuals in most evolutionary scenarios). The list could be continued, but the moral is clear. Simulation offers at best an impoverished version of the real-world arena, and a version impoverished in some dangerous ways: ways that threaten to distort our image of the operation of the agents by obscuring the contributions of environmental features and of real physical bodies.

For all that, the benefits of a judicious use of simulation can be large, especially when investigating evolutionary change. Large simulated populations are cheap to produce and easy to monitor. Fitness evaluation can be automated relative to behavior within the virtual environment. Real-world engineering problems are completely bypassed. In addition, large-scale simulated evolution offers vast time savings in comparison with the use of repeated real-world runs and evaluations.

For practical purposes, then, a mixed strategy seems to be indicated. Thus, theorists such as Nolfi, Miglino, and Parisi (1994) and Yamuchi and Beer (1994) use simulations for initial research and development and then transfer the results into real mobile robots. Of course, neural network controllers evolved to guide a simulated robot will hardly ever transfer without problems to a real-world system. But the simulation phase can at least be used to achieve rough settings for a variety of parameters, which can then be further tuned and adapted in the real-world setting.[13]

Finally, it should be noted that even pure simulation-based research can be immensely valuable, insofar as it allows the investigation of general issues concerning (e.g.) the interplay between individual learning and evolutionary change (Ackley and Littman 1992; Nolfi and Parisi 1991) and the properties of large collectives of very simple agents (Resnick 1994). As a means of understanding the detailed dynamics of real agent-environment interactions, however, simulations must always be taken with a large pinch of salt.

5.6 Understanding Evolved, Embodied, Embedded Agents

The process of natural design, it seems, will routinely outrun the imaginings of human theorists. In particular, biological evolution cares nothing for our neat demarcation between the merely physical and the computational or informational. Gross physical features such as mechanical and electrical loss, friction, and noise can all be exploited alongside familiar computational strategies (e.g., neural network learning) so as to yield robust solutions to problems of surviving and responding. Moreover, as we have seen repeatedly in previous chapters, the environment can be actively exploited so as to transform the nature of the problems we confront. And, as was remarked in section 5.2, biological evolution must often tinker with old resources so as to yield new capacities: cognitive innovation is thus seldom made from whole, ideally engineered cloth. These factors conspire to render biological design curiously opaque. Take a videocassette recorder apart and you find a well-demarcated set of modules and circuit boards, each of which plays a delimited and specific role in yielding successful performance. That's because human designers (unsurprisingly) opt for the kind of overall componential design that makes most sense to serial, conscious reflection. The human brain appears to involve much less transparent kinds of componential structure and wiring, including vast amounts of recurrent circuitry which allow mutual and iterated modifications between many areas. And the role of the brain, in any case, is merely to get the body to go through the right motions. Adaptive success finally accrues not to brains but to brain-body coalitions embedded in ecologically realistic environments. A large and currently unresolved question therefore looms: How are we to study and understand (not just replicate) the adaptive success of biological creatures—creatures whose design principles do not respect the intuitive boundaries between cognition, body, and world?

One possibility, currently gaining ground, is to replace the standard cognitive-scientific tools of computational theorizing and representation talk with those of Dynamical Systems theory. The argument goes like this: The image of cognition as the generation of computational transformation of internal representations is (it is said) a throwback to the idea of the brain as, in essence, the seat of a fundamentally *disembodied* kind of intelligence. It is a throwback because representations, thus conceived, are supposed to stand in for real-world items and events, and reasoning is supposed to occur in a kind of inner symbolic arena. But real embodied intelligence, we have seen, is fundamentally a means of *engaging* with the world—of using active strategies that leave much of the information out in the world, and cannily using iterated, real-time sequences of body-world interactions to solve problems in a robust and flexible way. The image here is of two coupled complex systems (the agent and the environment) whose joint activity solves the problem. In such cases, it may make little sense to speak of one system's *representing* the other.

The idea can be elusive, so an example may help. Tim van Gelder invites us to consider, in this light, the operation of the Watt governor described in section 5.5. The governor, recall, maintains the flywheel at a constant speed by using two weighted arms which swing out so as to close the throttle valve as speed of rotation increases and open it as speed decreases. Van Gelder (1995, p. 348) contrasts this with the operation of an imaginary "computational governor" that would operate as follows:

Measure the speed of the flywheel.
Compare the actual speed against the desired speed.
If there is a discrepancy, then
  measure the current steam pressure,
  calculate the desired alteration in steam pressure,
  calculate the necessary throttle-valve adjustment.
Make the throttle-valve adjustments. Return to step 1.

The computational governor thus uses explicit measurements of speed and steam pressure, which are fed into further processes for calculating the necessary adjustments. The Watt governor, in contrast, folds the stages of measurement, computation and control into a single process involving the reciprocal influences of the speed and angle of the arm and the speed of the engine. The best way to understand the operation of the Watt governor, van Gelder notes, is to think not in terms of representations and computations but in terms of feedback loops and closely coupled physical systems. Such phenomena are the province of standard Dynamical Systems theory. Let us pause to make its acquaintance.

Dynamical Systems theory is a well-established framework[14] for describing and understanding the behavior of complex systems (see, e.g., Abraham and Shaw 1992). The core ideas behind a Dynamical Systems perspective are the idea of a state space, the idea of a trajectory or a set of possible trajectories through that space, and the use of mathematics (either continuous or discrete) to describe the laws that determine the shapes of these trajectories.

The Dynamical Systems perspective thus builds in the idea of the evolution of system states over time as a fundamental feature of the analysis. As a general formalism it is applicable to all existing computational systems (connectionist as well as classicist), but it is also more general, and it can be applied to the analysis of noncognitive and noncomputational physical systems as well.

The goal of a Dynamical Systems analysis is to present a picture of a state space whose dimensionality is of arbitrary size (depending on the number of relevant system parameters), and to promote an understanding of system behaviors in terms of location and motion within that abstract geometric space. To help secure such an understanding, a variety of further constructs are regularly invoked. These constructs capture the distinctive properties of certain points or regions (sets of points) in the space as determined by the governing mathematics. The mathematics typically specifies a dynamical law that determines how the values of a set of state variables evolve through time. (Such a law may consist, for example, in a set of differential equations.) Given an initial state, the temporal sequence of states determined by the dynamical law constitutes one trajectory through the space. The set of all the trajectories passing through each point is called the *flow*, and its shape is the typical object of study. To help understand the shape of the flow, a number of constructs are used, including that of an attractor (a point or a region) in the space such that the laws governing motion through the space guarantee that any trajectory passing close to that region will be "sucked into" it. Related concepts include the "basin of attraction" (the area in which an attractor exerts its influence) and "bifurcations" (cases where a small change in the parameter values can reshape the flow, yielding a new "phase portrait"—i.e., a new depiction of the overall structure of basins and boundaries between basins).

The Dynamical Systems approach thus provides a set of mathematical and conceptual tools that support an essentially geometric understanding of the space of possible system behaviors. To get the flavor of these tools in use, consider once again the work on evolving insect leg controllers described in section 5.4. In attempting to understand the operation of a single leg controller,[15] Beer (1995b) highlights the role of a systematic flipping between two fixed-point attractors. The first comes into play when a foot has just been put down and a "stance phase" has begun. The evolution of this state takes the system close to a fixed-point attractor. As the leg continues to move, however, this attractor disappears to be replaced by a second attractor elsewhere in the state space, toward which the system state then evolves. This second attractor corresponds to a "swing phase." The switch between these fixed points is due to a set of bifurcations that occur as the leg moves through a certain angle. The effect of this is to switch the phase portrait of the controller between the two fixed-point attractors. Should the leg-angle sensor be disabled, the dynamics collapses to a fixed point, freezing the insect into a permanent stance phase. Notice especially that the dynamics Beer describes belong not to the neural network controller *per se* but rather to the coupled system comprising the controller and the insect body (leg). It is the reciprocal interplay of controller and leg (mediated by the leg sensor's angle-detecting capacity) that yields the state-space trajectory just described.

This kind of geometric, state-space-based understanding is, to be sure, both valuable and informative. It remains an open question, however, to what extent such explanations can replace, rather than merely complement, more traditional understandings couched in terms of computational transitions and inner representational states. The radical position (which predicts the wholesale replacement of computation and representation talk by geometric Dynamical Systems talk) faces two crucial challenges.

The first challenge concerns scaling and tractability. Even the 30-neuron leg controller constitutes a dynamical system of such complexity that our intuitive geometric understanding breaks down. Moreover, the detailed mathematics of Dynamical Systems theory becomes steadily less tractable as the number of parameters and the size of the state space increase. As a result, Beer's analysis was in fact conducted only for a simpler, five-neuron system controlling a single leg. The practical applicability of Dynamical Systems theory to highly complex, high-dimensional, coupled systems (like the human brain) must therefore be in serious doubt.

The second and more fundamental challenge concerns the type of understanding such analyses provide. This type of understanding threatens to constitute abstract description rather than full explanation. We learn what the system does and when it does it, and what patterns of temporal evolution its behavior displays; but this understanding, although valuable, does not seem to be exhaustive. In particular, we are often left—as I will later argue in detail—with an impoverished understanding of the *adaptive role* of components, and of the internal functional organization of the system.

The best aspects of the dynamical analyses, I suggest, are their intrinsic temporal focus and their easy capacity to criss-cross brain/body/

environment boundaries. I shall highlight the temporal issues in a subsequent chapter. The boundary issue should already be clear: By treating the brain as a dynamical system, we treat it in essentially the same terms as we treat bodily mechanics and environmental processes. As a result, it becomes especially easy and natural to characterize adaptive behavior in terms of complex couplings of brains, bodies, and environment.

I propose, therefore, to argue for a somewhat ecumenical stance. The tools of Dynamical Systems theory are a valuable asset for understanding the kinds of highly environmentally coupled behaviors I have highlighted. But they should be treated as *complementary* to the search for computational and representational understandings. The case for complementarity will occupy us for the next several chapters.

6 Emergence and Explanation

6.1 Different Strokes?

What kind of tools are required to make sense of real-time, embodied, embedded cognition? In particular, is there a range of emergent phenomena that depend so closely on the coupling of brain, body, and world that traditional analyses are bound to fail? I shall argue that emergent phenomena *do* demand some new modes of explanation and study, but that these new modes are best seen as complementary to (not in competition with) more familiar analytic approaches. Certainly, we will see an increasing sensitivity to what might be termed the ecological[1] determination of the roles of various inner states and processes (i.e., the way what needs to be internally represented and computed is informed by the organism's location in, and interactions with, a wider environment). And certainly, we will see the flip side of this same sensitivity: increased attention to the overall dynamics of whole organism/environment systems. But neither of these developments compromises our need to understand the contribution of neurophysiologically real components to the psychologically characterized abilities of an agent—a project that still appears to require the use of some quite traditional analytic tools. A successful cognitive science, I shall argue, will thus study both the larger dynamics of agent/environment systems and the computational and representational microdynamics of real neural circuitry.

6.2 From Parts to Wholes

In this section I distinguish three styles of cognitive scientific explanation. The styles are quite general, and they cross-classify particular programming styles (such as connectionist vs. classicist).

Componential Explanation

To explain the functioning of a complex whole by detailing the individual roles and the overall organization of its parts is to engage in componential[2] explanation. This is the natural explanatory style to adopt when, for example, we explain the workings of a car, a television set, or a washing machine. We explain the capacities of the overall system by adverting to the capacities and roles of its components, and the way they interrelate.

Componential explanation, thus construed, is the contemporary analogue to good old-fashioned *reductionistic* explanation. I avoid the vocabulary of reduction for two reasons. First, much of the philosophical discussion about reduction assumed that reduction named a relation between theories, and that theories were linguaform, law-involving constructs. But in many cases (especially biology and artificial intelligence) what we might otherwise naturally think of as reductive explanations do not take this form. Instead, they involve the development of partial models which specify components and their modes of interaction and which explain some high-level phenomena (e.g. being a television receiver) by adverting to a description of lower-level components and interactions.[3] These are reductionist explanations in a broader sense—one which "componential explanation" seems to capture. My second reason is that to contrast emergent explanation with reductionist explanation would be to invite a common misunderstanding of the notion of emergence—viz., to suggest that emergentist accounts embrace mystery and fail to explain how higher-level properties arise from basic structures and interactions. Recent emergentist hypotheses are by no means silent on such matters. The contrast lies in the *ways* in which the lower-level properties and features combine to yield the target phenomena. This kind of emergentist explanation is really a special case of reductionist explanation, at least as intuitively construed, since the explanations aim to render the presence of the higher-level properties unmysterious by reference to a multitude of lower-level organizational facts.[4] For these reasons, then, it will be more accurate and less confusing to contrast emergent explanation with componential explanation than with reductionist theorizing in general.

Modular programming methods in classical AI[5] lent themselves quite nicely to a componential form of explanation. In attempting to understand the success of such a program, it is often fruitful to isolate the various subroutines, modules, etc. and to display their role in dividing the target problem into a manageable series of subproblems (Dennett 1978a).

Recent "connectionist" work, as Wheeler (1994) points out, is likewise amenable to a kind of componential explanation. Solutions to complex problems such as the recognition of handwritten Zip codes (Le Cun et al. 1989) exploit highly structured, multi-layer networks (or networks of networks). In such cases it is possible to advance our understanding of how the system succeeds by asking after the roles of these gross components (layers or subnets). This kind of explanation is most compelling when the components admit of straightforward representational interpretation—that is, when the target systems have reliably identifiable internal configurations of parts that can be usefully interpreted as "representing aspects of the domain . . . and reliably identifiable internal components that can be usefully interpreted as algorithmically transforming those representations" (Beer 1995a, p. 225). In short: there is a relation between the componential analysis of intelligent systems and the image of such systems as trading in internal representations, for the distinctive roles of the posited components are usually defined by reference to the form or content of the internal representations they process.

"Catch and Toss" Explanation

This is my pet name for an approach that takes seriously many of the insights of embodied, embedded cognition but continues to view them through the lens of a traditional analysis. The main characteristic of the "catch and toss" mode is that the environment is still treated as just a source of *inputs* to the real thinking system, the brain. The concession to the embodied perspective involves recognizing that such inputs can lead to actions that simplify subsequent computations. The traditional image of an input-thought-action cycle is maintained, but the complex and reciprocal influences of real-world action taking and internal computation are recognized. Research on animate vision displays something of this character in its description of how low-resolution visual input can lead to real-world actions (such as moving the head or the fovea) that in turn generate more input suitable for higher-resolution processing. Here we confront a description that recognizes the multiple and complex ways in which the inner jobs can be altered and simplified by means of real-world structure, bodily dynamics, and active interventions in the world. But we also find a quite traditional emphasis on, and concern with, the realms of inner processing, internal representations, and computations (such as the construction of minimal internal databases encoding special-purpose "indexical" representations such as "My coffee cup is yellow"; see Ballard 1991, pp. 71–80). The peaceful coexistence of these two images (of the active, embedded system and of the primacy of the inner processing economy) is maintained by a firm insistence on the boundary between the brain and the world. The world tosses inputs to the brain, which catches them and tosses actions back. The actions may alter or simplify subsequent computations, by causing the world to toss back more easily usable inputs and so on. In short, there is a strong commitment to *interactive* modes of explanation, but the traditional focus on representation and computation in the individual brain is respected. One reason for this, implicit in the "catch and toss" idea itself, is that much of the focus in these cases is on simple feedback chains in which the system's actions alter its next inputs, which control the next action and so on. In such cases the relative low dimensionality of the interactions allows us to understand the system's behavior using quite conventional tools. By contrast, as the complexity and the dimensionality of crucial interactions increase, it becomes difficult (perhaps impossible) to conceptualize the situation by simply superimposing a notion of feedback loops on top of our standard understanding. Such critical complexity arises when the number of feeedback processes increases and when the temporal staging of the various processes goes "out of synch," allowing feedback to occur along multiple channels and on multiple, asynchronous time scales.[6]

Emergent Explanation

Emergent explanation is at once the most radical and the most elusive member of our trinity. Whereas "catch and toss" explanation is really just a sensitive and canny version of componential explanation, emergent explanation aims to offer a whole new perspective on adaptive success. At the heart of this new perspective lies, of course, the tricky idea of emergence itself. Let's approach it warily, by way of a sequence of illustrative examples.

J. A. Scott Kelso, in his excellent treatment *Dynamic Patterns* (1995), presents the classic example of fluid heated from below. Specifically, he describes the behavior of cooking oil heated in a pan. When the heat is first applied, there is little temperature difference between the top and the bottom of the oil, and we observe no motion in the liquid. However, as the temperature increases, the body of oil begins to move in a coordinated fashion—we observe what Kelso (ibid., p. 7) describes as "an orderly, rolling motion." The source of this motion is the temperature difference between the cooler oil at the top and at the hotter oil at the bottom. The hotter and less dense oil rises, while the cooler and denser oil falls—a cycle which then repeats as the old cool oil, now at the bottom, gets hotter and rises, only to cool once again, and so on. The result is the persisting rolling motion known as *convection rolls*. The appearance of convection rolls is an example of an emergent self-organizing property of a collection of molecules, not unlike the self-organization of slime-mold cells described in chapter 4. Kelso (ibid., pp. 7–8) comments:

> The resulting convection rolls are what physicists call a collective or cooperative effect, which arises without any external instructions. The temperature gradient is called a control parameter in the language of dynamical systems. Note that the control parameter does not prescribe or contain the code for the emerging pattern. It simply leads the system through a variety of possible patterns or states. . . . Such spontaneous pattern formation is exactly what we mean by self-organization: the system organized itself, but there is no 'self,' no agent inside the system doing the organizing.

The idea, of course, is not that the emergent patterns are totally uncaused—obviously the proximal cause is the application of heat to the pan. Rather, it is that the observed patterns are largely explained by the collective behavior (under specified conditions) of a large ensemble of simple components (the molecules), none of which is playing a special or leading role in controlling or orchestrating the process of pattern formation. In fact, once the rolling motion begins, it feeds and maintains itself in a way that is characteristic of "self-organizing" systems. These systems are such that it is simultaneously true to say that the actions of the parts cause the overall behavior and that the overall behavior guides the action of the parts. For a homely example of this idea (sometimes called "circular causation"), consider the way the actions of individuals in a crowd combine to initiate a rush in one direction, and the way that activity then sucks in and molds the activity of undecided individuals and maintains and reinforces the direction of collective motion. Such phenomena invite understanding in terms of *collective variables*—variables that fix on higher-level features which are crucial to the explanation of some phenomenon but which do not track properties of simple components. Instead, such variables may track properties that depend on the interaction of multiple components—properties such as the temperature and pressure of a gas, the rate of acceleration of a panicking crowd, or the amplitude of the convection rolls formed in a heated liquid. By plotting the values of such collective variables as a system unfolds over time, we may come to understand important facts about the actual and possible behavior of the larger system. And by plotting relations between the values of the collective variables and control parameters (such as the temperature gradient) we may come to understand important facts about the circumstances in which such higher-level patterns will emerge, when one higher-level pattern will give way to another, and so on.

One basic sense of the elusive term "emergence" is thus to hand. There is emergence whenever interesting, non-centrally-controlled behavior ensues as a result of the interactions of multiple simple components within a system. We have already, however, rubbed shoulders with a second sense of emergence—one rooted primarily in ideas of organism-environment interactions. This kind of emergence, which characterizes a lot of the work in real-world robotics described in previous chapters, can be illustrated with a simple example drawn from Steels 1994. Steels invites us to imagine a robotic agent that needs to position itself between two poles so as to recharge itself. The charging station is indicated by a light source. One (non-emergentist) solution would be to endow the robot with sensors that measure its position relative to the poles and with a subroutine that computes a trajectory between the poles. An alternative (emergentist) solution relies on two simple behavior systems whose environmental interactions yield positioning between the poles as a kind of side effect. The behavior systems are (1) a phototaxis system that yields a zigzag approach to any light source and (2) an obstacle-avoidance system that causes the robot to turn away when it hits something. With these two simple systems in place, the target behavior emerges smoothly and robustly. The robot is attracted to the light and zigzags toward it. If it touches a pole, it retreats, but it is then attracted by the light and tries again, this time from a new angle. After a few tries, it finds the only position in which its behavior systems are in equilibrium—a position near the light, but not touching either pole. The pole-orienting behavior counts as emergent because no actual component computes a pole-centering trajectory—instead, phototaxis, obstacle avoidance, and local environmental structure (the location of the light source) collectively bring about the desired result. We thus confront a second sense of emergence—one that turns on the idea of functionally valuable side effects brought about by the interaction of heterogeneous components, and which foregrounds the notion of interactions between behavior systems and local environmental structure. The two senses of emergence thus correspond roughly to what (in section 4.2) I termed the distinction between direct and indirect forms of emergence.

Let us now try to go a little further and clarify the common theme uniting the various cases. Is there, in short, a reasonably precise, nontrivial account of the overarching idea of an emergent feature?

Sometimes the general notion of emergence is equated with the idea of unexpected behaviors. (There are traces of this in Steel's emphasis on "side effects," although he is aware of the dangers and tries hard to avoid them.) The trouble here is that what is unexpected to one person may be just what someone else predicts—a canny engineer might design the pole-centering robot precisely so as to exploit the interactions between basic components and the world as a way of solving the charging problem. Yet the solution itself retains the characteristic flavor of emergence, even if the outcome was predicted from the start. What we really need, then, is an observer-independent criterion—or, at least, a criterion less hostage to the vagueries of individual expectations.

A more promising idea, also mentioned by Steels, equates emergent phenomena with ones that require description in a new vocabulary: a vocabulary quite different from the one we use to characterize the powers and properties of the components themselves. Steels gives the example of chemical properties such as temperature and pressure, which do not figure in descriptions of the motion of individual molecules but which are needed to describe the behavior of aggregates of such items. This looks promising, but it still won't quite do. The reason is that the vocabulary switch also characterizes cases that are not, intuitively, cases of real emergence. A hi-fi system comprising an amplifier, a tuner, and speakers exhibits behavior some of which is best described in a vocabulary that does not apply to any of the individual components, yet such a system looks like a prime candidate for good old-fashioned componential explanation.[7]

A better account of emergence (for our purposes, at any rate) is a generalization of a distinction between what Steels (ibid.) calls *controlled* variables (which track behaviors or properties that can be simply and directly manipulated) and *uncontrolled* variables (which track behaviors or properties that arise from the interaction of multiple parameters and hence tend to resist direct and simple manipulation). Consider Douglas Hofstadter's story about an operating system that begins to "thrash around" once about 35 users are on line. In such a case, Hofstadter notes, it would be a mistake to go to the system's programmer and ask to have the "thrashing number" increased to, say, 60. The reason is that the number 35 is not determined by a simple inner variable upon which the programmer can directly act. Instead, "that number 35 emerges dynamically from a host of strategic decisions made by the designers of the operating system and the computer's hardware and so on. It is not available for twiddling." (Hofstadter 1985, p. 642) Here we have a fully *systems-internal* version of an uncontrolled variable. In other cases, changing the variable might require adjusting a host of both inner and outer (environmental) parameters whose collective behavior fixes the variable's value. Emergent phenomena, on this account, are thus any phenomena whose roots involve uncontrolled variables (in this extended sense) and are thus the products of collective activity rather than of single components or dedicated control systems. Emergent phenomena, thus understood, are neither rare nor breathtaking: nonetheless, getting target behaviors to arise as functions of uncontrolled variables has not been a common strategy in AI, and such behaviors, when they arise, demand types of understanding and explanation that go beyond both the componential model and the interactive model rehearsed above.

Two final examples will help. The first, drawn from Resnick 1994b, concerns a strategy for getting simulated termites to collect wood chips and gather them into piles. One solution would be to program the termites to take chips to predesignated spots. Relative to such a solution, chip piling would count as a controlled variable, as piling behavior would be under direct control and would be fully "twiddleable." An emergentist solution, in contrast, yields the behavior indirectly via the combined effects of two simple rules and a restricted environment. The rules are these: "If you are not carrying anything and you bump into another wood chip, pick it up; If you are carrying a wood chip and you bump into another wood chip, put down the wood chip you are carrying." (ibid., p. 234) It is not obvious that such a strategy will work, as it allows chips to be removed from piles as easily as they can be added! Nonetheless, 2000 scattered chips, after 20,000 iterations, become organized into just 34 piles. The piling behavior ends up overwhelming the de-piling behavior because whenever (by chance) the last chip is removed from an incipient pile, that location is effectively blocked; under the two rules, no new pile can ever begin there. Over time, then, the number of possible pile locations in the artificial grid diminishes, forcing the chips to congregate in the remaining locations. It is the unprogrammed and environmentally determined feature of "location blocking" that enables piling activity to outrun de-piling activity. In this example, it is clear that piling behavior is not directly controlled but rather emerges from the interplay between the simple rules and the restricted environment.

A second example: Hallam and Malcolm (1994) describe a simple solution to the problem of getting a robot to follow walls. You build into the robot a bias to veer to the right, and locate on its right side a sensor which is activated by contact and which causes the device to turn a little to the left. Such a robot will, on encountering a wall on the right, first move away (thanks to the sensor) and then quickly veer back to reencounter the wall (thanks to the bias). The cycle will repeat, and the robot will follow the wall by, in effect, repeatedly bouncing off it. In fact, as Tim Smithers has usefully pointed out in a personal communication, this solution requires a quite delicate balance between amount of "right veer" and amount of "left turning." Smithers also points out that this general idea of using "opposing forces to achieve stable regulated behavior" can be seen in early water-clock technology—a nice case of emergent timekeeping! The point to notice, in any case, is that the wall-following behavior described above emerges from the interaction between the robot and its environment. It is not subserved by any internal state encoding a goal of wall following. We, as external theorists, lay on the wall-following description as a gloss on the overall embedded behavior of the device. In both of these cases, Steel's distinction between controlled and uncontrolled variables seems to give us what we need. The story can also be successfully applied to pole centering and, I suspect, to any cases of indirect emergence. Nonetheless, the emphasis on phenomena that cannot be controlled or manipulated by altering the values of a single parameter fails to encompass intuitively emergent phenomena such as the appearance of convection rolls in heated liquids. The reason is that the convection rolls are under the control of a simple parameter (the temperature gradient, or—to move to proximal causes—the applied heat) and can be effectively "twiddled" (to use Hofstadter's memorable phrase) as a result. In fact, the temperature gradient that drives the motion is called a control parameter precisely because it governs the collective behaviors of the system in such a powerful way.

Given this important class of cases, I think a better account of emergence (a kind of weak generalization of the idea of an uncontrolled variable) is simply this: a phenomenon is emergent if it is best understood by attention to the changing values of a collective variable. Some quick points about this definition:

- A collective variable is a variable that tracks a pattern resulting from the interactions among multiple elements in a system (section 6.2 above; Kelso 1995, pp. 7, 8, 44). Thus, all uncontrolled variables are collective variables.
- To accommodate cases of indirect emergence, we extend the relevant notion of the "system" to include (at times) aspects of the external environment, as in the case of the pole-centering robot.
- Different degrees of emergence can now be identified according to the complexity of the interactions involved. Multiple, nonlinear,[8] temporally asynchronous interactions yield the strongest forms of emergence; systems that exhibit only simple linear interactions with very limited feedback do not generally require understanding in terms of collective variables and emergent properties at all.
- Phenomena may be emergent even if they are under the control of some simple parameter, just so long as the role of the parameter is merely to lead the system through a sequence of states themselves best described by appeal to a collective variable (e.g., the temperature gradient that leads

the liquid through a sequence of states described by a collective variable marking the varying amplitude of the convection rolls—see Kelso 1995, p. 8).

• Emergence, thus defined, is linked to the notion of what variables figure in a good explanation of the behavior of a system. This is a weakly observer-dependent notion, since it turns on the idea of a good theoretical account and hence builds in some relation to the minds of the human scientists. But at least it does not depend on the vagaries of individual expectations about system behavior.

6.3 Dynamical Systems and Emergent Explanation

What is the most effective explanatory framework for understanding emergent phenomena? A widely shared negative intuition is that classical componential explanation, at least, often fares badly in such cases (Steels 1994; Maes 1994; Wheeler 1994). There are two rather distinct reasons for such failure.

One reason turns on the fact that many (not all) emergent cognitive phenomena are rooted in factors that spread across an organism and its environment. In such cases (and we saw several examples above) we ideally require an explanatory framework that (1) is well suited to modeling both organismic and environmental parameters and (2) models them both in a uniform vocabulary and framework, thus facilitating an understanding of the complex interactions between the two. A framework that invokes computationally characterized, information-processing homunculi is not, on the face of it, an ideal means of satisfying these demands.

A second reason turns on the nature of components. When each of the components makes a distinctive contribution to the ability of a system to display some target property, componential analysis is a powerful tool. But some systems are highly homogeneous at the component level, with most of the interesting properties dependent solely upon the aggregate effects of simple interactions among the parts. One example (van Gelder 1991; Bechtel and Richardson 1992) would be a simple connectionist network in which the processing units are all markedly similar and the interesting abilities are largely attributable to the organization (by weighted, dense connectivity) of those component parts. A more complex case occurs when a system is highly nonhomogeneous, yet the contributions of the parts are highly inter-defined—that is, the role of a component C at time t_1 is determined by (and helps determine) the roles of the other components at t_1, and may even contribute quite differently at a time t_2, courtesy of complex (and often nonlinear—see note 8) feedback and feedforward links to other subsystems. Thus, even internal nonhomogeneity and on-line functional specialization is no guarantee that a componential analysis will constitute the most revealing description.

These complexities are reflected in Wimsatt's (1986) lovely description of "aggregate systems." Aggregate systems are the ones for which componential explanation is best suited. Such systems are defined as ones in which the parts display their explanatory relevant behavior even in isolation from one another, and in which the properties of a small number of subsystems can be invoked to explain interesting systemic phenomena.[9] As the complexities of interaction between parts increases, the explanatory burden increasingly falls not on the parts but on their organization. At such a time, we are driven to seek new kinds of explanatory frameworks. As we shall later see, it is likely that advanced biological cognition falls somewhere near the middle of this continuum. The systems have distinct and functionally specialized neural components, but the complex and often nonlinear interactions (feedback and feedforward relations) between these components may be crucial determinants of most intuitively "psychological" phenomena. Good explanations, in such cases, require both a traditional componential explanation and something else. But what else?

Given our two desiderata (viz., that we accommodate both organism-environment interactions and complex interactions between components), it is natural to consider the framework (briefly introduced in chapter 5 above) of Dynamical Systems theory—a theoretical approach that provides a set of tools for describing the evolution of system states over time (Abraham and Shaw 1992). In such descriptions, the theorist specifies a set of parameters whose collective evolution is governed by a set of (usually) differential equations. One key feature of such explanations is that they are easily capable of spanning organism and environment. In such cases the two sources of variance (the organism and the environment) are treated as coupled systems whose mutual evolution is described by a specific set of interlocking equations. The behavior of a wall-mounted pendulum placed in the environmental setting of a second such pendulum provides an easy example. The behavior of a single pendulum can be described using simple equations and theoretical constructs such as attractors and limit cycles,[10] but two pendulums placed in physical proximity tend, surprisingly, to become swing-synchronized over time. This synchronization admits of an elegant Dynamical Systems explanation treating the two pendulums as a single coupled system in which the motion equation for each pendulum includes a term representing the influence of the other's current state, the coupling being achieved via vibrations running through the wall.[11] Most important in the present context, Dynamical Systems theory also provides a new kind of explanatory framework. At the heart of this framework is the idea of explaining a system's behavior by isolating and displaying a set of variables (collective variables, control parameters, and the like) which underlie the distinctive patterns that emerge as the system unfolds over time and by describing those patterns of actual and potential unfolding in the distinctive and mathematically precise terminology of attractors, bifurcation points, phase portraits, and so forth. (See section 5.6.)

There are many ways in which a typical Dynamical Systems explanation varies from a traditional, component-centered understanding. The most puzzling difference, at first sight, is that Dynamical Systems theory seems to want to explain behaviors by describing behaviors. Yet (intuitively, at least), the provision of even a rich and detailed description appears to fall well short of the provision of an explanation, which typically reduces puzzlement by revealing something of the hidden mechanisms that bring a behavior about. In addition, many scientists and philosophers believe that certain physical systems (such as the brain) depend on special organizational principles and hence require a vocabulary and an explanatory style very different from those used to explain the coordination of pendula or the dripping of taps. Again, Dynamical Systems theory surprises us by using the same basic approach to tackle many superficially very different kinds of real-world phenomena. This helps to explain why many cognitive scientists, on encountering this style of explanation, are disappointed to find detailed stories about patterns in gross behavior and little by way of "real mechanism." This is indeed a surprise if you are expecting a special kind of story focused on hidden, inner events. But the cognitively motivated Dynamical Systems theorist believes that both neural dynamics and gross bodily dynamics flow from the same deep underlying principles of self-organization in complex systems. It is natural, working from within such a perspective, to treat both kinds of patterns in similar ways—as Kelso (1995, p. 28) puts it, "the claim on the floor is that both overt behavior and brain behavior, properly construed, obey the same principles."[12]

To get the true flavor of the kinds of explanation at issue here, let us turn to a real case study due to Kelso et al. (1981) and nicely summarized in chapter 2 of Kelso 1995. The study centers on the phenomenon of rhythmical behavior, and in particular on the production of rhythmic finger motions. Try to move your two index fingers from side to side so that they are moving at the same frequency. You will probably find that this can be achieved either by moving the two fingers so that the equivalent muscles of each hand contract at the same time or by ensuring that the equivalent muscles are exactly out of phase (one contracting as the other expands). The same two stable strategies describe the behavior of car windshield wipers: typically, the wipers move in phase. But a few models are set up to display slightly unnerving anti-phase coordination. The important difference is that human subjects can settle into either mode, according to how they begin the action sequence. Moreover, the anti-phase strategy is stable only at low frequencies of oscillation. If a subject begins in anti-phase mode and is then asked to steadily increase the speed of oscillation, there occurs, around a certain critical frequency, an abrupt shift or phase transition. In a striking example of spontaneous pattern alteration, the anti-phase twiddling gives way to phased twiddling. (The same kind of spontaneous alteration occurs when a horse switches, at a certain velocity, from a trot to a canter. These two styles of locomotion involve quite different inter-limb coordination strategies—see Kelso 1995, pp. 42–43.)

How should we explain this pattern of results? Kelso set out to do so by first investigating which variables and control parameters would best describe the behaviors. The crucial variable, he discovered, was one that tracked the phase relationship between the fingers. This variable, as we saw, is constant for a wide range of individual finger oscillation frequencies, and changes suddenly when the frequency hits a critical value. It is a collective variable since it cannot be defined for a single component (finger) but only for the larger system. Frequency of movement is thus the control parameter for the phase relation which is now plotted as a collective variable. The real meat of the analysis then lies in the provision of a detailed mathematical description of the system thus described—a set of equations displaying the space of possible temporal evolutions of relative phase as governed by the control parameter. Such a description effectively describes the state space (see chapter 5) of the system showing, among other things, which areas in the space act as attractors (variable values toward which the system will tend from certain other locations in the space). Haken et al. (1985) found just such a description and were able to display the detailed patterns of coordination corresponding to different values of the control parameter. Important features of the model included not only its ability to describe the observed phase transitions without positing any "switching mechanism" above and beyond the collective dynamics but also its ability to reproduce the results of minor interference with the system, such as occurs if one finger is briefly forced out of its stable phase relation. The model of Haken et al. also generated accurate predictions of features such as the time taken to switch from out of phase to in phase.[13]

It should now be clearer why the dynamical account is not merely a nice description of the observed phenomena. It owes its status as an *explanation* to its ability to illuminate what philosophers call counterfactuals: to inform us not just about the actual observed behavior of the system but also about how it will behave in various other circumstances. Nonetheless, these explanations still lack one powerful feature of their more traditional cousins. They are not constrained to constitute detailed recipes for building the kinds of devices they both describe and explain. In this, they differ from familiar models in which a behavior is explained by showing how it arises from the properties of a variety of well-understood components. Traditional computational models, for example, have the very real virtue of decomposing complex tasks into sequences of simpler and simpler ones to the point where we see how to carry them out given only the basic resources of logic gates, memory boards, and so forth.

On the positive side, Dynamical Systems explanations, with their apparatus of collective variables and coupled behaviors, are naturally suited to spanning multiple interacting components and even whole agent-environment systems. Whereas the standard framework seems geared to describing computation and representation in agent-side processing, the Dynamical Systems constructs apply as easily to environmental features (e.g., the rhythms of a dripping tap) as to internal information-processing events. It is this easy ability to describe larger integrated systems that leads theorists such as Beer and Gallagher (1992) and Wheeler (1994) to prefer Dynamical Systems theory over classical componential approaches for the explanation of emergent, often environment-involving types of behaviors. Behaviors so far studied tend to be relatively basic ones, such as legged locomotion (see chapter 5 above) and visually guided motion. But the intuition of many theorists is that the bulk of everyday biological intelligence is rooted in canny couplings between organisms and specific task environments, and thus that this style of explanation may extend well beyond accounts of relatively "low-level" phenomena. Indeed, Port and van Gelder 1995 contains several examples of Dynamical Systems theorizing applied to such high-level tasks as planning, decision making, language production, and event recognition.

It is important to remember, however, that the system parameters tracked in these Dynamical Systems explanations can be arbitrarily far removed from facts about the real internal structure and processing of the agent. Van Gelder (1991) notes that a Dynamical Systems story tracking the behavior of a car engine over time might need to fix on a parameter, such as temperature, that does not correspond to any internal component or to any directly controlled variable. This can occur, van Gelder notes, because "in its pure form, dynamical explanation makes no reference to the actual structure of the mechanism whose behavior it is explaining. It tells us how the values of the parameters of the system evolve over time, not what it is about the way the system itself is constituted that causes those parameters to evolve in the specified fashion. It is concerned to explore the topographical structure of the dynamics of the system, but this is a wholly different structure than that of the system itself." (ibid., p. 500)

Intermediate options are clearly also available. Salzman (1995) offers a Dynamical Systems explanation of how we coordinate multiple muscles in speech production. He notes that the coordinative dynamics must be specified in abstract informational terms, which do not directly track either biomechanical or neuroanatomical structure. Instead, "the abstract dynamics are defined in coordinates that represent the configurations of different constriction types e.g. the bilabial constrictions used in producing /b/, /p/, or /m/, the alveolar constrictions used in producing /d/, /t/, or /n/ etc." (ibid., p. 274). These constriction types are defined in physical terms involving items such as lip aperture and lip protrusion. But the Dynamical Systems story is defined over the more abstract types mentioned above. This is an intermediate case insofar as it is clear how the more abstract parameters cited in the Dynamical Systems analysis are related to physical structures and components of the system.

Such intermediate analyses are of great importance. Cognitive science, I shall next argue, cannot afford to do without any of the various explanatory styles just reviewed, and it is therefore crucial that we ensure that the various explanations somehow interlock and inform one another. I shall now develop an argument for this explanatory liberalism and show how the requirement of explanatory interlock imposes powerful additional constraints on our theorizing.

6.4 Of Mathematicians and Engineers

Just how powerful is the pure Dynamical Systems style of explanation and analysis? My view, as will become more increasingly clear over the next few chapters, is that it provides a crucial part of the understanding we need, but that (at least at present) it cannot take us all the way. To see why, we must first be clear about what I mean by a pure Dynamical Systems style of explanation.

A pure Dynamical Systems account will be one in which the theorist simply seeks to isolate the parameters, collective variables, and so on that give the greatest grip on the way the system unfolds over time—including (importantly) the way it will respond in new, not-yet-encountered circumstances. The pure Dynamical Systems theorist is thus seeking mathematical or geometrical models that give a powerful purchase on the observable phenomena. This is good science, and it is explanatory science (not mere description). Moreover, as we just saw, much of the distinctive power and attractiveness of these approaches lies in the way they can fix on collective variables—variables whose physical roots involve the interactions of multiple systems (often spread across brain, body, and world). But this distinctive power comes at a cost: these "pure" models do not speak directly to the interests of the engineer. The engineer wants to know how to build systems that would exhibit mind-like properties, and, in particular, how the overall dynamics so nicely displayed by the pure accounts actually arise as a result of the microdynamics of various components and subsystems. Such a person may well insist that a full understanding of the working system will positively require pure dynamical stories such as those just rehearsed. However, he or she will not think such stories sufficient to constitute an understanding of how the system works, because they are pitched at such a distance from facts concerning the capacities of familiar and well-understood physical components. By contrast, a standard computational account (connectionist or classical) is much closer to a recipe for actually building a device able to exhibit the target behaviors. This is because all the basic state transitions involved in the specification are constrained to be reproducible by known combinations of basic operations that can be performed using logic gates, connectionist processing units, or whatever.

In a certain sense, what is accomplished by a pure dynamical discussion is more closely akin to a sophisticated *task analysis* than to a fully worked out computational story. But it is a task analysis that is both *counterfactually pregnant* (see section 6.3) and potentially *wide*. It is wide insofar as it can "fold together" aspects of the problem space that depend on the external environment as well as those that depend on properties of the individual organism. In such cases there will be multiple ways of implementing the dynamics described, some of which may even divide subtasks differently among body, brain, and world. For example, body fat may do for infant A what artificial weights do for infant B, and complex computations may do for creature C what compliance in elastic muscles does for creature D. Identical gross dynamics may thus emerge from very different "divisions of labor."

The complaint, then, is that commanding a good pure dynamical characterization of the system falls too far short of possessing a recipe for building a system that would exhibit the behaviors concerned. One response to this complaint (a response I have often heard from diehard fans of Dynamical Systems theory) is to attack the criterion itself. Why should we insist that real understanding requires "knowing how to build one"? Esther Thelen (personal communication) notes that "by such a criterion, we would need to throw out nearly all biology"—not to mention economics, astronomy, geology, and who knows what else. Why should cognitive science buy into an explanatory criterion so much more demanding than those proper to just about every science?

Despite its surface plausibility, this reply really misses the point. It does so by taking the claim about buildability just a bit too literally. What is really being suggested is not that *in fact* we should be able to build systems that exhibit the desired features (although AI, to its credit, often aims to do just that), but that we should understand something of how the larger-scale properties are rooted in the interactions of the parts. Perhaps we cannot build our own volcanos, but we do understand how subterranean forces conspire to create them. We may, in addition, seek powerful accounts of the ebb and flow of volcanic activity over time, and we may even do so by isolating control parameters, defining collective variables, and so on. A full understanding of the nature of volcanic activity depends, no doubt, on the simultaneous pursuit and careful interlocking of both types of explanation. In the relevant sense, then, we do know how to build volcanos, whirlwinds, solar systems, and all the rest! Our problems in actually carrying out the construction stem from practical difficulties (of scale, materials, etc.) and not from any lack of the requisite level of understanding.

The buildability criterion thus needs softening to allow for the large number of cases in which other problems stand in our way. Typical snags, to quote from an aptly titled paper[14] by Fred Dretske, might be: "The raw materials are not available. You can't afford them. You are too clumsy or not strong enough. The police won't let you." (Dretske 1994, p. 468) Conversely, Dretske notes, the mere fact that you can build something does not guarantee that you really understand it—we can all assemble a kit and be none the wiser. The core (and I believe correct) claim is, therefore, just that to really understand a complex phenomenon it is at least *necessary* that we understand at least something of how it is rooted in the more basic properties of its biologically or physically proper parts. What this ultimately requires, I suggest, is continually probing beyond the level of collective variables and the like so as to understand the deeper roots of the collective dynamics themselves.

The good news is that, occasional rhetoric aside, most of the proponents of a dynamical approach recognize and respond to this very need. Thelen and Smith (1994), having described the embodied, embedded behavior of infants in great detail, go on to pursue issues concerning the dynamics of the underlying neural organizations. As they themselves point out, their depiction of the changing dynamical landscapes of the infants (the shifting attractors) leaves them "completely uninformed about the more precise mechanisms of changing attractor stability" (ibid., p. 129). In response to this need, they then pursue a Dynamical Systems approach at the level of neural organization. Kelso (1995, p. 66) is, if anything, even clearer, insisting that a "tripartite scheme" involving a minimum of three levels (the task or goal level . . . , collective variable level, and component level) is required to provide a complete understanding." Kelso also notes, importantly, that what actually counts as a component or a collective variable will depend in part on our specific explanatory interests. To use his own example, nonlinear oscillators may be treated as components for some purposes. Yet the nonlinear oscillatory behavior is itself a collective effect that arises from the interactions of other, more fundamental parts.

Randall Beer, in his careful and progressive attempts to understand the operation of neural-network controllers for simple model agents, stresses the need to understand the detailed dynamics of individual neurons, of coupled pairs of neurons, of coupled pairs of neurons coupled to simple bodies, and so on up the scale. In short, Beer seeks a Dynamical Systems understanding that will go all the way down, and one relative to which the special properties of ever larger and more complex systems should begin to make more sense. (See, e.g., Beer 1995.) Common to all these theorists, then, is a recognition that the explanatory aspirations of cognitive science go beyond the careful depiction of embodied, embedded behavior, and beyond even the genuine explanations of such behavior that can be given in terms of collective variables keyed to making sense of gross observed behavior. What ultimately distinguishes these approaches from more traditional work is an insistence (Kelso et al.) or a suspicion (Beer) that the familiar notions of internal representation, information processing, and (perhaps) computation do not provide the best vocabulary or framework in which to understand the remaining issues concerning neural organization. Instead, these authors are betting on the use of a Dynamical

Systems vocabulary to describe and explain all levels of biological organization. My view, as the next few chapters will make clear, is that we will not only need a mix of levels of analysis (something like Kelso's "tripartite scheme") but also a mix of explanatory tools, combining Dynamical Systems constructs with ideas about representation, computation, and the information-processing role of distinguishable subcomponents. To get the broad flavor of such a mixed approach, consider a concrete example.

6.5 Decisions, Decisions

Componential explanation and "catch and toss" explanation are both well suited to explaining adaptive behavior by unraveling the contributions of specific agent-side components. "Catch and toss" explanation differs largely in its explicit recognition of the profound differences which attention to environmental opportunities and the demands of real-time action can make to our hypotheses concerning the internal information-processing organization required. The pure Dynamical Systems approach to explaining emergent phenomena, in contrast, looks to import a whole new perspective, one which focuses on the evolution of overall system parameters and which is especially well suited to modeling the complex interplay between multiple agent-side parameters and environmental ones. Thus described, it seems almost obvious that both types of explanation (the information-processing, componential-style analysis and the global-dynamics-style analysis) are needed and should be required to interlock gracefully. Yet several recent writings suggest an alternative, more imperialist point of view. Dynamical Systems theory, they suggest, is to be *preferred over* talk of information-processing decompositions and internal components that traffic in representations. Such a radical view can be sustained only by adopting an unduly impoverished vision of the goals of cognitive science.

Consider the goal of explaining the systematic effects of various kinds of local damage and disruption. The stress on gross system parameters, which helps us understand the dynamics that obtain within well-functioning organism-and-environment systems, must often obscure the details of how various inner systems contribute to that coupling, and thus how the failure of such systems would affect overall behavior. Yet an important body of work in cognitive neuroscience aims precisely to plot the inner organization that explains patterns of breakdown after local damage (Farah 1990; Damasio and Damasio 1994). Such explanations typically adopt both a modular/componential perspective and a representation-invoking perspective. This kind of understanding *complements* any broader understanding of global dynamics. Each explanatory style helps capture a distinct range of phenomena and helps provide different types of generalization and prediction.

For example, Busemeyer and Townsend (1995) present an elegant application of Dynamical Systems theorizing to understanding decision making. The framework they develop, called Decision Field Theory, describes how preference states evolve over time. They describe dynamical equations which plot the interplay of various gross factors (such as the long-term and short-term anticipated value of different choices) and which also predict and explain the oscillations between likely choices that occur during deliberation. These are explained as effects of varying how much attention the decision maker is currently giving to various factors. The account captures and explains several interesting phenomena, including the apparent inconsistencies between preference orderings measured by choice and those measured by selling price.[15] A whole class of generalizations, explanations, and predictions thus falls out of the specific equations they use to model the evolution of the chosen parameters and variables over time.

Other kinds of explanation and generalization, however, are not subsumed by this level of description. Thus consider the famous mid-nineteenth-century case of Phineas Gage, a railroad construction foreman who suffered a terrible injury when a tamping iron was thrust right through his face, skull, and brain. Amazingly, Gage survived and regained all his logical, spatial, and physical skills. His memory and intelligence were not affected, yet his life and personality changed dramatically. He was no longer trustworthy, or caring, or able to fulfill his duties and commitments. The damage to his brain had caused, it seemed, a very specific yet strange effect—it was almost as if his "moral centers" had been destroyed (Damasio et al. 1994). More accurate, it seemed that his ability to "make rational decisions in personal and social matters" (ibid.) had been selectively compromised, leaving the rest of his intelligence and skills intact.

In recent years, a team of neuroscientists specializing in brain imaging analyzed Gage's skull and were able, using computer-aided simulations, to identify the probable sites of neural damage. By identifying specific neural structures as the site of the damage, Damasio et al. (ibid.) were able to begin to make sense of Gage's selective disturbances (and those of others—see the case of E.V.R. in Damasio et al. 1990). The damage was to the ventromedial regions of both frontal lobes—areas that appear to play a major role in emotional processing. This finding led the Damasio team to speculate on a special role for emotional responses in social decision making.[16] Partly inspired by such case studies, the Damasios also developed a more general framework for the explanation of selective psychological deficits. This is the "convergence-zone hypothesis" explained in some detail in the next chapter. A distinctive feature of this hypothesis, as we shall see, is the way it combines attention to the basic functional compartmentalization of the brain with recognition of the role of larger-scale integrative circuitry. A full account of deficits such as Gage's and E.V.R.'s thus requires, it seems, a combination of some quite familiar kinds of information-processing localization (assigning distinct tasks to different areas of sensory and motor cortex) and the kinds of larger-scale analysis that implicate multiple areas linked together by complex webs of feedback and feedforward connectivity.

The details of this story will become clearer once we turn up the focus on contemporary neuroscience (in chapter 7). The point, for current purposes, is not to assess the details of any such proposal. It is merely to note that the proposal of Damasio et al. aims for a type of understanding that is not present in the global depiction offered by Decision Field Theory, which is quite clearly not designed to either predict or illuminate the kind of unexpectedly selective disturbance to the decision-making process that these neuroanatomy-motivated studies address. This is not a criticism of DFT, which itself provides a type of understanding, prediction, and explanation which the Damasio proposal does not. This is because DFT is free to treat emergent properties of the whole, intact, well-functioning system as collective variables, and hence provides a vocabulary and a level of analysis well suited to capturing patterns in the temporally evolving behavior of intact well-functioning agents. Moreover, it is these more abstract descriptions that will often serve us best if we seek to understand the couplings between whole systems and their environments. Globally emergent features, we should readily concede, often play an important role in illuminating accounts of such couplings. Since the two styles of explanation are naturally complementary, there is, however, no need for the kind of competition that some of the fans of Dynamical Systems analysis seem to encourage. Instead, we should clearly distinguish two explanatory projects, each having its own associated class of generalizations. One project aims to understand the way intact agents and environments interrelate, and in so doing it may invoke abstract, globally emergent parameters. The other seeks to understand the specific information-processing roles of various inner subsystems in producing behavior, and hence it helps explain whole classes of phenomena (e.g., the effects of local damage) which its counterpart simply does not address.

Indeed, one natural way to think of the two projects just outlined is to depict the componential analysis as providing (in part) a story about the detailed implementation of the more global and abstract Dynamical Systems story. Van Gelder (1991) is skeptical about the value of such implementational stories, at least as regards the understanding of complex neural networks; he notes (p. 502) that componential (or, as he says, "systematic") explanation is of little help in cases where "the 'parts' of the structure are so many and so similar, and key parameters . . . do not refer to parts of the system at all." But while this may be true for understanding the behaviors of single, relatively homogeneous connectionist networks, it seems manifestly untrue as regards the brains of most biological organisms. A more realistic picture, I suggest, must countenance three equally important and interlocking types of explanation and description:

(1) An account of the gross behaviors of the well-functioning organism in the environment—an account that may invoke collective variables whose componential roots span brain, body, and world.

(2) An account that identifies the various components whose collective properties are targeted by the explanations proper to (1). Two important subtasks here are to identify relevant neural components and to account for how these components interact.

(3) An account of the varying information-processing roles played by the components (both internal and external) identified in (2)—an account that may well assign specific computational roles and representational capacities to distinct neural subsystems.

Satisfying explanations of embodied, embedded adaptive success must, I claim, touch all three bases. Moreover, each type of explanation imposes constraints and requirements on the others. There can be no legitimate collective variables in (1) that lack microdynamic implementation detail in (2), and such detail cannot be fully understood without the gross systems-level commentary on the roles of the various components provided by (3). The best way to achieve this, it would seem, is to pursue all three of the explanatory types isolated earlier in my discussion: componential analysis, so as to assign broad information-processing roles to neural structures; "catch and toss" analysis, to track the way the organism acts on and is acted on by the environment; and emergentist analysis, to describe the classes of adaptive behavior that depend most heavily on collective variables and organism-environment interactions.

6.6 The Brain Bites Back

A full account of embodied, embedded, and emergence-laden cognition must, it seems, do justice to several kinds of data. One such body of data concerns changes in a system's gross behavior over time. Another concerns, e.g., the specific effects of local, internal damage to the system. To explain such heterogeneous phenomena, the theorist should be willing to exploit multiple kinds of explanatory tools, ranging from analyses that criss-cross the organism and the environment, to ones that quantify over multiple inner components and complex connectivities, to ones that isolate components and offer a functional and representational commentary on their basic roles.

Emergent properties will figure in this explanatory activity at two levels. First, there will be internally emergent features: features tracked by collective variables constituted by the interaction of multiple inner sources of variation. Second, there will be behaviorally emergent features: features tracked by collective variables constituted by interactions between whole, functioning organisms and the local environment. Both classes of emergent property need to be understood, and Dynamical Systems theory provides a set of tools that can help in each arena. But these multiple explanatory endeavors are not autonomous. Collective variables must be cashed out in real (neural and environmental) sources of variance. And basic componential specializations must be identified and factored into our understanding and models. Failure to do the latter will result in explanatory failure farther down the line—for example, when we are confronted by data concerning the selective impairments caused by local brain damage. In the next chapter, we will begin flesh out this broad framework by taking a closer look at some recent neuroscientific research.

7

The Neuroscientific Image

7.1 Brains: Why Bother?

Does cognitive science really need to bother with the biological brain? To the casual observer, the answer will seem obvious: of course it must—how *else* can we hope to achieve a better understanding of the mind? What's more, the casual observer is right! It is all the more surprising, then, that influential research programs in cognitive science have so often downplayed or ignored neuroscientific studies in their attempts to model and explain mental phenomena. One popular reason for such inattention was the claim, common among early workers in symbolic artificial intelligence, that the right level of description of the physical device (for psychological purposes) lay at a fair remove from descriptions of neuronal structures and processes. Instead, it was believed that some much more abstract level of description was required—for example, a description in terms of information-processing roles in a computational system.[1] The fine details of neuronal organization, it was thought, constituted one specific solution to the problem of how to physically construct a device that would satisfy such an abstract computational story—but that was all.[2]

With the advent (or rebirth) of connectionist models, all that began to change. These models were deliberately specified in a way that reduced the distance between the computational story and the broad nature of neuronal implementation. The detailed fit between connectionist work and real brain theory was often far flimsier than one might have hoped. But, as connectionism matured, attempts to further bridge the gap[3] became increasingly common, and a real synthesis of the computational and neuroscientific perspectives looked to be in the cards.

But connectionist research was also being pulled in another direction: the direction, highlighted throughout this book, of attending to the details of embodied and environmentally embedded cognition. This emerging emphasis, I claim, should not be allowed to swamp attempts to develop increasingly neurally plausible models. Indeed, the two perspectives need to proceed hand in hand. We should indeed view the brain as a complex system whose adaptive properties emerge only relative to a crucial backdrop of bodily and environmental structures and processes. However, to fully understand these extended processes we must understand in detail the contributions of specific neural systems and the complex interactions among them. The stress on organism-environment interactions should thus not be seen as yet another excuse for cognitive science to avoid confrontation with the biological brain.

The real question, then, is not "should we study the brain?" but "how should we study the brain?" What *kind* of neuroscientific models best engage with our emphases on embodied action and real-time success? And if such models exist, how well are they supported by neuroanatomical and cognitive neuroscientific data and experiments? I shall suggest that the most promising class of neuroscientific models has three main characteristics:

(1) the use of multiple, partial representations,
(2) a primary emphasis on sensory and motor skills,
and
(3) a decentralized vision of the overall neural economy.

In the following sections I will sketch and discuss some examples of neuroscientific conjectures of this broad type, and indicate some lines of continuity with research into embodied, embedded cognition.

7.2 The Monkey's Fingers

Consider a simple-sounding question: How does the monkey's brain guide the monkey's fingers? For many years neuroscientists endorsed a simple and intuitive picture. Part of the monkey's brain, the story went, was the site of a somatotopic map: a region in which groups of spatially clustered neurons were dedicated to the control of individual digits. Activity in a group would cause the correlated finger to move. To move several fingers at once required simultaneous activity in several neuronal groups. This vision of how the monkey's brain controls its fingers was immortalized in the "homuncular" images of the spatial subdivisions of the Ml (motor area 1) brain region, which depicted distinct neuronal groups controlling each individual digit arranged in lateromedial sequence.

The model is neat and intuitive, and it would represent a nice solution to problems that require fingers to move independent of one another (as in skilled piano playing). But it is not nature's solution, as more recent research demonstrates. To see this, note an obvious prediction of the simple homuncular model. The model predicts that movements involving several digits should demand the activation of more, and more widespread, neurons than movements of an isolated digit. In addition, the model predicts that thumb movements should be accompanied by activity in the more lateral region of the Ml hand area, with movements of the other digits following in sequence until the most medial region (corresponding to the movements of the little finger) is reached. Neither prediction is borne out. Marc Schieber and Lyndon Hibbard, neuroscientists at Washington University School of Medicine, found that movements of each individual digit were accompanied by activity spread *throughout* the Ml hand area. In addition, it has been observed that more motor cortex activity is required for precise movements than for more basic whole hand movements, and that some motor-cortex neurons seem devoted to the *prevention* of movements of other digits when a target digit must act in isolation.

Schieber (1990, p. 444) suggests that we make sense of all this by thinking of isolated digit movements as the complex case, with "more rudimentary synergies, such as those used to open and close the whole hand" as the basic adaptation. Such an adaptation makes perfect sense for a creature whose primary need is to grasp branches and swing, and the overall diagnosis fits our evolutionary perspective on natural cognitive design to a T. The basic problem is one of producing fast, fluent, and environmentally appropriate action. Neural coding strategies are selected to facilitate a particular range of time-critical grasping behaviors. This basic, historically determined need shapes the solutions to more recent problems involving isolated digit movements (as in piano playing). To achieve these more recent goals, Schieber (ibid.) suggests that "the motor cortex might then superimpose control in part on the phylogenetically older subcortical centers and in part directly on . . . spinal motorneurons, so as to adjust the movement of all the different fingers." Evolution thus tinkers with whole-hand synergies, devoting neural resources to the suppression of movements as much as to their production, so as to enable precise movements.

The moral, it seems, is that biological evolution can select internal coding schemes that look alien and clumsy at first sight but which actually represent quite elegant solutions to the combined problem of serving basic needs and making the most of existing resources. More generally, the neuroscientific literature abounds with cases of somewhat unexpected neural codings. For example, some neurons in the posterior parietal cortex of the rat have been found to respond maximally (in the context of running a radial maze) to specific combinations of head orientation and the presence of some local landmark or feature (McNaughton and Nadel 1990, pp. 49–50), others to specific turning motions made by the rat (another case of the kind of agent-oriented, motocentric representations featured in previous chapters).

Schieber's model also illustrates the role, in natural cognition, of distributed internal representations. This topic has loomed large in recent work on artificial neural networks.[4] A distributed representation is an inner encoding in which the target content is not carried by an individual resource (e.g. a single neuron) and is not necessarily carried by a spatially localized group of units or neurons. Instead, the content (concerning, e.g., the motion of an individual digit) is carried by a pattern of activation which is spread across a population of neurons or units. Distributed encodings present a number of advantages and opportunities. For example, the pattern itself can encode significant structural information in such a way that minor variations in the pattern reflect small but sometimes important differences in what is currently represented. And it becomes possible to use methods of overlapping storage so that each individual neuron can play a role in encoding many different things (just as the number 2 can appear in many different numerical patterns: "2, 4, 6," "2, 3, 4," "2, 4, 8," and so on). When this kind of overlapping storage is systematically exploited, so that semantically related items are represented by overlapping but nonidentical patterns, there follow further advantages of generalization (new items or events can be given nonarbitrary codings based on the extent to which the new item resembles an old one) and graceful degradation (limited physical damage is less troublesome since multiple elements will participate in the coding for each broad class of item or event, and performance will be sensible as long as some are spared). These advantages are discussed in detail elsewhere (Clark 1989, 1993); the point for now is simply that the brain may be using quite complex, overlapping, spatially distributed representational schemes even in cases where we might intuitively have expected a simple, spatially localized encoding strategy, as in the case of the M1 motor area.[5] Nature's way, it seems, is to use spatially overlapping distributed encodings to govern related (but nonidentical) types of finger movement. The final picture is thus one in which specific cortical neurons play a role in controlling *several* finger muscles and do so by participating in widely spatially extended patterns of activity which correspond to different types and directions of finger movements.

7.3 Primate Vision: From Feature Detection to Tuned Filters[6]

To get more of the flavor of current neuroscientific research, let us take a brief foray into the increasingly well-understood world of primate vision. We already saw, in the computational work on animate vision, the likely extent to which a thrifty nature may rely on cheap cues and local environmental state so as to minimize internal computational load. But even when we allow for this, the complexity of the mechanisms of real vision remains staggering. What follows is a necessarily truncated sketch based on the recent work of David Van Essen, a leading researcher on primate vision who (conveniently for me) is based at the Washington University School of Medicine.[7]

Neuroanatomical research has uncovered a multiplicity of anatomical parts and pathways that seem to play special roles in visual processing. Cognitive neuroscience aims, using a wide diversity of experimental and theoretical tools, to identify the different response characteristics of participating neurons and neuronal populations. Anatomically, the macaque monkey possesses at least 32 visual brain areas and over 300 connecting pathways. Major areas include early cortical processing sites such as V1 and V2, intermediate sites such as V4 and MT, and higher sites such as IT (inferotemporal cortex) and PP (posterior parietal cortex) (plate 1). The connecting pathways tend to go both ways—e.g. from V1 to V2 and back again. In addition, there is some "sideways" connectivity—e.g. between subareas within V1.

Felleman and Van Essen (1991) describe the overall system as comprising ten levels of cortical processing. Some of the most important of these will now be described. Subcortically, the system takes input from three populations of cells, including so-called magnocellular and parvocellular populations. One subsequent processing pathway is largely concerned with the magnocellular input, another with parvocellular input. This division makes sense in view of the different types of low-level information each population "specializes" in. Parvo (P) cells have high spatial and low temporal resolution; magno (M) cells have high temporal resolution. As a result, M cells enable the perception of rapid motions, whereas P cells underpin (among other things) color discrimination. Selective destruction of P cells prevents a monkey from distinguishing color but leaves motion recognition intact.

The magno-denominated (MD) stream of processing includes many populations of neurons sensitive to the direction of a stimulus motion, especially in area MT. Electrical stimulation of part of MT can cause a monkey to "perceive" left motion when the target object is in fact moving to the right (Salzman and Newsome 1994). At still higher stages in the processing hierarchy (such as MSDT) there is evidence for cells sensitive to quite complex motion stimuli, such as spiral motion patterns (Graziano et al. 1994). The MD stream is ultimately connected to the posterior parietal cortex, which appears to use spatial information to control such high-level functions as deciding *where* objects are and planning eye movements.

Meanwhile, the question of what things are (object recognition) is left to a different stream of processing: one rooted especially in parvocellular inputs, proceeding through V1, V4, and PIT (posterior inferotemporal areas), and leading into central and anterior inferotemporal areas. This pathway seems to specialize in form and color. By the level of V4, there is evidence of cells sensitive to quite complex forms such as concentric, radial, spiral, and hyperbolic stimuli (plate 2). Higher still, individual cells in the inferotemporal cortex respond maximally to complex geometrical stimuli such as faces and hands. But (and this is crucial) these maximal responses do not exhaustively specify a cell's information-processing role. Although a cell may respond maximally to (e.g.) a spiral pattern, the same cell will respond to some degree to multiple other patterns also. It is often the tuning of a cell to a whole set of stimuli that is most revealing. This overall tuning enables one cell to participate in a large number of distributed patterns of encoding, contributing information both by its being active and by its degree of activity. Such considerations lead Van Essen and others to treat cells not as simple feature detectors signaling the presence or absence of some fixed parameter but rather as *filters* tuned along several stimulus dimensions, so that differences in firing rate allow one cell to encode multiple types of information.[8] There is also strong evidence that the responses of cells in the middle and upper levels of the processing hierarchy are dependent on attention and other shifting parameters (Motter 1994), and that even cells in V1 have their response characteristics modulated by the effects of local context (Knierim and Van Essen 1992). Treating neurons as tunable and modulable filters provides a powerful framework in which to formulate and understand such complex profiles. The basic image here invoked is also consonant with the design perspective advocated by Tim Smithers (see section 5.5 above), in which very simple sensory systems are themselves analyzed as tuned filters rather than as simple feature-detecting channels.

Recent work on primate vision thus shows an increasing awareness of the complexity and sophistication of biological coding schemes and processing pathways. Yet this increasing appreciation of both complexity and interactive dynamics does not render the primate visual system analytically opaque. Instead, we see how the system progressively separates, filters, and routes information so as to make various types of information available to various components (e.g. to inferotemporal cortex and to posterior parietal cortex) and so as to allow both low-level and higher-level visual cues to guide behavior as and when required. A full understanding of (e.g.) animate-vision strategies (recall chapter 1) will thus require both an appreciation of many kinds of complex internal dynamics and an understanding of how the embodied, embedded cognizer *uses* such resources to exploit environmental features and locally effective cues in the service of adaptive success.

7.4 Neural Control Hypotheses

An important development in recent cognitive neuroscience involves the increasing recognition of the role of neural control structures. Neural control structures, as I shall use the term, are any neural circuits, structures, or processes whose primary role is to modulate the activity of other neural circuits, structures, or processes—that is to say, any items or processes whose role is to control the inner economy rather than to track external states of affairs or to directly control bodily activity. A useful analogy, suggested by Van Essen et al. (1994), is with the division of processes in a modern factory, where many processes are devoted not to the actual construction of the product but rather to the internal trafficking of materials. Likewise, many neuroscientists now believe, large amounts of neural capacity are devoted to the trafficking and handling of information. The role of certain neuronal populations, on those accounts, is to modulate the flow of activity between other populations so as to promote certain classes of attentional effect, multi-modal recall, and so forth.

Van Essen et al. (1994) posit, in this vein, neural mechanisms devoted to regulating the flow of information between cortical areas. Such regulation allows us, they suggest, to (e.g.) direct an internal window of visual attention at a specific target (such as a letter of the alphabet which appears at some random location in the visual field), or to direct the very same motion command to any one of a number of different body parts. In each case, the computational expense of generating a distinct signal for each case would be prohibitive. By developing a single resource which can be flexibly "targeted" on various locations, vast computational savings are effected. The key to such flexible targeting, Van Essen et al. argue, is the use of populations of "control neurons" which dynamically route information around the brain. Nor do they leave this proposal at the level of the intuitive sketch presented above. Instead, they develop a detailed neural-network-style model of the operation of such controllers and relate the model to a variety of known neurological substrates and mechanisms. The highly context-dependent response profiles of some cortical neurons (mentioned in section 7.3) may itself be explained by the operation of these mechanisms for the routing and rerouting of incoming information.

Another kind of neural control hypothesis is based on the idea of "reentrant processing" (Edelman and Mountcastle 1978; Edelman 1987). It is well known that the brain includes many pathways which link distant cortical areas and which lead back from higher to lower brain areas. The idea of reentrant processing is that these "sideways and descending" pathways are used to control and coordinate activity in multiple (often much lower-level) sites. The pathways carry "reentrant signals" which cause the receiving sites to become active. Consider two populations of neurons, each of which is responding to different types of external stimuli (e.g. from vision and touch), but which are reciprocally interconnected by such reentrant pathways. The reciprocal pathways would allow the goings-on at one site to become usefully correlated, over time, with goings-on at the other site. Such correlations could come to encode higher-level properties, such as the combinations of texture and color distinctive of some particular class of objects.

As a final example of a neural control hypothesis, consider the recent attempt by Damasio and Damasio (1994) to develop a general framework for the explanation of selective psychological deficits. Selective deficits occur when, usually as a result of a brain lesion or trauma, an individual loses specific types of cognitive capacity while others are left relatively intact. For example, the patient known as Boswell was found to be selectively impaired in the retrieval of knowledge concerning unique entities (e.g. specific individuals) and events (e.g. specific episodes of autobiography, unique places, unique objects) (Damasio et al. 1989). Nonetheless, his more general categorical knowledge remained intact. He could identify items as cars, houses, people, etc. Boswell showed no deficits of attention or perception, and his ability to acquire and display physical skills was not affected.

Damasio and Damasio (1994) describe a framework capable of accounting for such patterns of deficits. The key feature of their proposal is that the brain exploits "convergence zones": areas that "direct the simultaneous activation of anatomically separate regions whose conjunction defines an entity" (ibid., p. 65). A convergence zone is then defined as a neuronal grouping in which multiple feedback and feedforward loops make contact. It is a region in which several long-range corticocortical feedback and feedforward connections converge. The function of a convergence zone is to allow the system to generate (by sending signals back down to multiple cortical areas involved in early processing) patterns of activity across widely separated neuronal groups. When we access knowledge of concepts, entities, and events, the Damasios suggest, we exploit such higher-level signals to re-create widespread patterns of activity characteristic of the contents in question. If we suppose that different classes and types of knowledge require different complexes of co-activation, managed by different convergence zones, we can begin to see how local brain damage could selectively impair the retrieval of different types of knowledge. To explain the dissociation between knowledge of unique and non-unique events, however, we need to introduce the additional notion of a hierarchy of convergence zones. Recall that convergence zones, as imagined by the Damasios, project both backward (reactivating earlier cortical representations) and forward (to higher convergence zones). The higher zones can economically prompt widespread lower-level activity by exploiting feedback connections to the earlier links in a convergence zone hierarchy. The basic hypothesis is thus as follows:

> . . . the level at which knowledge is retrieved (e.g. superordinate, basic object, subordinate) would depend on the scope of multiregional activation. In turn, this would depend on the level of convergence zone that is activated. Low level convergence zones bind signals relative to entity categories. . . . Higher level convergence zones bind signals relative to more complex combinations. . . . The convergence zones capable of binding entities into events . . . are located at the top of the hierarchical streams, in anteriormost temporal and frontal regions. (ibid., p. 73)

The idea, then, is that retrieval of knowledge of unique entities and events requires the conjunctive activation of many more basic loci than does knowledge of non-unique entities and events (the former subsuming the latter, but not vice versa). Similarly, knowledge of concepts would require the conjunctive activation of several distinct areas, whereas knowledge of simple features (e.g. color) may be restricted to a single area. Assuming a hierarchy of convergence zones extended in neural space, this picture would explain why damage to early visual cortices selectively impairs knowledge of simple features such as color, whereas damage to intermediate cortices affects knowledge of non-unique entities and events, and damage to anterior cortices impairs responses concerning unique individuals and events.

According to this framework, distinct but overlapping neural systems promote access to different types of knowledge. The more complex the conjunctions of information required to fix a class of knowledge, the more such coordinative activity is needed. This, in turn, implicates correlatively higher sites in a hierarchy of convergence zones which correspond to increasingly anterior loci within the temporal cortices. Damasio and Damasio stress that they are not depicting the damaged regions as the physical sites of different classes of knowledge. Rather, the damaged regions are the control zones that promote the conjunctive activation of several quite distant areas. These are typically early sensory and motor cortices, which would be led to re-create their proprietary responses to certain external stimuli by the reentrant signals. Summing up their suggestions, Damasio and Damasio comment that

> the (picture) we favor implies a relative functional compartmentalization for the normal brain. One large set of systems in early sensory cortices and motor cortices would be the base for "sense" and "action" knowledge. . . . Another set of systems in higher-order cortices would orchestrate time-locked activities in the former, that is, would promote and establish temporal correspondence among separate areas. (ibid., p. 70)

On this account, there are localized neural regions for several types of sensory and motor information, and for several levels of convergence-zone-mediated control. Higher-level capacities (such as grasping concepts) are, however, depicted as depending on the activity of *multiple* basic areas (in sensory and motor cortices) mediated by the activity of *multiple* convergence zones. Much of the explanatory apparatus for explaining phenomena like concept possession will thus require resources that go beyond the simple componential analyses introduced in chapter 6. We will need models that are especially well adapted to revealing the principles underlying phenomena that emerge from the complex, temporarily time-locked, coevolving activity of multiple components linked by multiple feedback and feedforward pathways. Classical componential analyses have not tended to fare well in such cases, and there thus looks to be a clear opening here for a Dynamical Systems account of the detailed implementation of the convergence-zone hypothesis. At the same time, the explanatory power of the theory is clearly tied up with the prior decomposition into basic processing areas (performing identifiable cognitive tasks) and into a well-defined set of convergence zones whose different switching activities likewise map onto different classes of knowledge retrieval. It is only in the light of this decomposition and functional commentary that the model can predict and explain the selective effects of local brain damage on retrieval of knowledge concerning unique entities and events. In this case the presence of the component-based analysis seems essential as a means of bridging the gap between the phenomena to be explained (viz., deficits affecting specific types of knowledge) and the models we create. Without talk of the cognitive functions of early sensory cortices, and without talk of higher-level cortical structures as being specialized to re-create specific complexes of cognitive activity, we would not understand why any additional descriptions of the detailed dynamics of interactions between components were actually explanatory of the psychological phenomena.

Notice, finally, that neural control hypotheses fall well short of depicting the brain as a centralized message-passing device, for there is an important distinction between imagining that some internal control system has access to all the information encoded in various subsystems, and imagining a system which can open and close channels connecting various subsystems.[9] All that the neural control hypotheses sketched above demand is the latter, channel-controlling capacity. They are thus a far cry from the traditional vision of a "central executive" system. The "higher centers" posited by the Damasios do not act as storehouses of knowledge "ftp'd" from the lower-level agencies. They are instead "merely the most distant convergence points from which divergent retroactivation can be triggered" (Damasio and Damasio 1994, p. 70). Much opposition to an information-processing approach is, I believe, more properly cast as opposition to a rich "message-passing" vision of mind (see, e.g., Brooks 1991). Thus, Maes (1994, p. 141) notes that work on adaptive autonomous agents eschews the use of classical modules that "rely on the 'central representation' as their means of interface." Instead, such researchers propose modules that interface via very simple messages whose content rarely exceeds signals for activation, suppression, or inhibition. As a result, there is no need for modules to share any representational format—each may encode information in highly proprietary and task-specific ways (ibid., p. 142). This vision of decentralized control and multiple representational formats is both biologically realistic and computationally attractive. But it is, as we have seen, fully compatible both with some degree of internal modular decomposition and with the use of information-processing styles of (partial) explanation.

Neural control hypotheses thus constitute a powerful mixture of radicalism and traditionalism. They are radical insofar as they offer a decentralized, non-message-passing model of higher cognition, insofar as they often depict higher cognition as arising from the time-locked activity of multiple, more basic types of sensorimotor processing area, and insofar as they recognize the complex, recurrent dynamics of neural processing. But they retain key elements of more traditional approaches, such as the use of an information-processing-style, component-based decomposition in which distinct neural components are associated with specific content-bearing roles.

7.5 Refining Representation

Contemporary neuroscience, as even this brief and sketchy sampling shows, displays an interesting mix of the radical and the traditional. It retains much of the traditional emphasis on componential and information-processing-based analyses of neural computation. But it does so within the broader context of a systemic understanding which is both increasingly decentralized and attentive to the role of complex recurrent dynamics. The notion of internal representation still plays a key role, but the image of such representations is undergoing some fundamental alterations—first because the question of what gets internally represented has been reopened both as a result of "bottom-up" studies of the response profiles of specific neural populations (as in Schieber's work on representation in the monkey's motor cortex) and as a result of increased awareness of the importance of the organism's ecological embedding in its natural environment (as in the animate-vision research reported earlier), and second because the question of *how* things are internally represented has been transfigured by connectionist work on distributed representation and by the recognition that individual neurons are best seen as filters tuned along multiple stimulus dimensions. This combination of decentralization, recurrence, ecological sensitivity, and distributed multidimensional representation constitutes an image of the representing brain that is far, far removed from the old idea of a single, symbolic inner code (or "Language of Thought"—see Fodor 1975 and Fodor 1986). It is representationalism and computationalism stripped of all excess baggage, and streamlined so as to complement the study of larger organism-environment dynamics stressed in previous chapters. To complete this project of integration and reconciliation, we next must now look more closely at the fundamental notions of computation and representation themselves.

8

Being, Computing, Representing

8.1 Ninety Percent of (Artificial) Life?

Ninety percent of life, according to Woody Allen, is just being there; and we have indeed charted lots of ways in which the facts of embodiment and environmental location bear substantial weight in explaining our adaptive success. The last two chapters, however, introduced some important caveats. In particular, we should not be too quick to reject the more traditional explanatory apparatuses of computation and representation. Minds may be essentially embodied and embedded and *still* depend crucially on brains which compute and represent. To make this ecumenical position stick, however, we need to specifically address some direct challenges concerning the very ideas of computation and representation (do they have nontrivial definitions compatible with new framework?) and some problems concerning the practical application of such notions in systems exhibiting emergent properties dependent upon processes of complex, continuous, reciprocal causation.[1]

8.2 What Is This Thing Called Representation?

Cognitive scientists often talk of both brains and computer models as housing "internal representations." This basic idea provided common ground even between the otherwise opposing camps of connectionism and classical artificial intelligence.[2] The differences between connectionists and the classicists concerned only the precise nature of the internal representational system, not its very existence. Classicists believed in a "chunky symbolic" inner economy in which mental contents were tokened as strings of symbols that could be read, copied, and moved by some kind of inner central processing unit. Connectionists believed in a much more implicit style of internal representation: one that replaced strings of chunky, manipulable symbols with complex numerical vectors and basic operations of pattern recognition and pattern transformation.

For all that, explicit, chunky symbolic representations and distributed vectorial connectionist representations were both seen as species of internal representation, properly so called. This overarching species is present, it has been argued, whenever a system meets certain intuitive requirements. Haugeland (1991) unpacks these by describing a system as (internal) representation using just in case:

(1) It must coordinate its behaviors with environmental features that are not always "reliably present to the system."
(2) It copes with such cases by having something else (in place of a signal directly received from the environment) "stand in" and guide bahavior in its stead.
(3) That "something else" is part of a more general representational scheme that allows the standing in to occur systematically and allows for a variety of related representational states (see Haugeland 1991, p. 62).

Point 1 rules out cases where there is no "stand-in" at all and where the environmental feature (via a "detectable signal") directly controls the behavior. Thus, "plants that track the sun with their leaves need not represent it or its position because the tracking can be guided directly by the sun itself" (ibid., p. 62). Point 2 identifies as a representation anything that "stands in" for the relevant environmental feature. But point 3 narrows the class to include only stand-ins that figure in a larger scheme of standing in, thus ruling out (e.g.) gastric juices as full representations of future food (ibid.). These conditions are on the right track. But the role of decouplability (the capacity to use the inner states to guide behavior in the absence of the environmental feature) is, I think, somewhat overplayed.

Consider a population of neurons in the posterior parietal cortex of the rat. These neurons carry information about the direction (left, right, forward) in which the animal's head is facing. They do so by utilizing a coding scheme that is "general" in something like the sense required by Haugeland's third condition. At least on my reading, the notion of a general representational scheme is quite liberal and does not require the presence of a classical combinational syntax in which items can be freely juxtaposed and concatenated. Instead, it requires only that we confront some kind of encoding *system*. And these may come in many, many varieties. For example, it will be sufficient if the system is such that items which are to be treated similarly become represented by encodings (such as patterns of activation in a population of neurons or an artificial neural net) which are close together in some suitable high-dimensional state space.[3] This kind of representational scheme is, in fact, characteristic of much of the connectionist work discussed earlier, and looks to characterize at least some of the systems of encoding found in biological brains. Populations of posterior parietal neurons in the rat are a case in point. Yet there is nothing in our description so far to suggest that these neurons can play their role in the absence of a continuous stream of proprioceptive signals from the rat's body. If such "decoupling" is not possible, we confront a case that nicely meets Haugeland's third condition (there is some kind of systematic coding scheme present) but not his other two (the coding scheme is not available to act as a stand-in in the absence of the incoming signals). What should we say of such a case?[4]

It seems reasonably clear that by glossing states of the neuronal population as codings for specific head positions we gain useful explanatory leverage. Such glosses help us understand the flow of information within the system, when, for example, we find other neuronal groups (such as motor-control populations) that consume the information encoded in the target population. The strict application of Haugeland's criteria would however, rule out the description of any such inner systems (of non-decouplable inner states) as genuinely representational. This seems unappealing in view of the very real explanatory leverage that the representational gloss provides, and it is also out of step with standard neuroscientific usage.

Haugeland's criteria thus seem a little too restrictive. It is important, however, to find some way of constraining the applicability of the idea of internal representation. For example, it is surely necessary to rule out cases of mere causal correlation and of overly simple environmental control. It is certainly true that the presence within a system of some kind of complex inner state is not sufficient to justify characterizing the system as representational. As Beer (1995a) and others have pointed out, all kinds of systems (including chemical fractionation towers) have complex internal states, and no one is tempted to treat them as representational devices. Nor is the mere existence of a reliable, and even nonaccidental, *correlation* between some inner state and some bodily or environmental parameter sufficient to establish representational status. The mere fact of correlation does not count so much as the nature and complexity of the correlation and the fact that a system in some way consumes[5] or exploits a whole body of such correlations for their specific semantic contents. It is thus important that the system uses the correlations in a way that suggests that the system of inner states *has the function* of carrying specific types of information.

There is a nice correlation between tides and the position of the moon; however, neither *represents* the other, since we do not find it plausible that (e.g.) the tides were selected, were designed, or evolved for the purpose of carrying information about the position of the moon. In contrast, it seems highly plausible that the population of neurons in the posterior parietal cortex of the rat is supposed (as a result of learning, evolution, or whatever—see Millikan 1984 and Dretske 1988) to carry information about the direction in which the animal's head is facing. And such a hypothesis is further supported when we see how other neural systems in the rat consume this information so as to help the rat run a radial maze.[6]

The status of an inner state as a representation thus depends not so much on its detailed nature (e.g., whether it is like a word in an inner language, or an image, or something else entirely) as on the role it plays within the system. It may be a static structure or a temporally extended process. It may be local or highly distributed. It may be very accurate or woefully inaccurate. What counts is that it is *supposed to* carry a certain type of information and that its role relative to other inner systems and relative to the production of behavior is precisely to bear such information. This point is nicely argued by Miller and Freyd , who add that "the strengths of representationalism have always been the basic normative conception of how internal representations should accurately register important external states and processes" and that "the weaknesses . . . have resulted from overly narrow assumptions about what sorts of things can function as representations and what sorts of things are worth representing" (1993, p. 13). To which (modulo the unnecessary emphasis on *accurate* registration) amen.

Keeping all these points in mind, let us call a processing story representationalist if it depicts whole systems of identifiable inner states (local or distributed) or processes (temporal sequences of such states) as having the function of bearing specific types of information about external or bodily states of affairs. Representationalist theorizing thus falls toward the upper reaches of a continuum of possibilities whose nonrepresentationalist lower bounds include mere causal correlations and very simple cases of what might be termed "adaptive hookup." Adaptive hookup goes beyond mere causal correlation insofar as it requires that the inner states of the system are supposed (by evolution, design, or learning) to coordinate its behaviors with specific environmental contingencies. But when the hookup is very simple (as in a sunflower, or a light-seeking robot), we gain little by treating the inner state as a representation. Representation talk gets its foothold, I suggest, when we confront inner states that, in addition, exhibit a systematic kind of coordination with a whole space of environmental contingencies. In such cases it is illuminating to think of the inner states as a kind of code that can express the various possibilities and which is effectively "read" by other inner systems that need to be informed about the features being tracked. Adaptive hookup thus phases gradually into genuine internal representation as the hookup's complexity and systematicity increase. At the far end of this continuum we find Haugeland's creatures that can deploy the inner codes in the total absence of their target environmental features. Such creatures are the most obvious representers of their world, and are the ones able to engage in complex imaginings, off-line reflection, and counterfactual reasoning. Problems that require such capacities for their solution are representation-hungry, in that they seem to cry out for the use of inner systemic features as stand-ins for external states of affairs. It is not, however, a foregone conclusion that creatures capable of solving such problems must use internal representations to do so. The cognitive scientific notion of internal representation brings with it a further commitment that we will gain explanatory leverage by treating identifiable inner substates or processes as the bearers of specific contents, and by deciphering the more general coding schemes in which they figure. Should such a project be blocked (e.g., should we find beings capable of thinking about the distal and the nonexistent whose capacities of reason and thought did not succumb to our best attempts to pin representational glosses on specific inner happenings), we would confront representing agents who did not trade in internal representations!

The question before us is, thus, as follows: What role, if any, will representational glosses on specific inner happenings play in the explanations of a mature cognitive science? To this question there exists a surprising variety of answers, including these:

(1) The glosses are not explanatorily important, but they may serve a kind of heuristic function.
(2) The glosses are actively misleading. It is theoretically misleading to associate specific inner states or processes with content-bearing roles.
(3) The glosses are part of the explanatory apparatus itself, and they reflect important truths concerning the roles of various states and processes.

Some (though by no means all) fans of Dynamical Systems theory and autonomous-agent research have begun to veer toward the most skeptical option—that is, toward the outright rejection of information-processing accounts that identify specific inner states or processes as playing specific content-bearing roles. The temptation is thus to endorse a radical thesis that can be summarized as follows:

Thesis of Radical Embodied Cognition Structured, symbolic, representational, and computational views of cognition are mistaken. Embodied cognition is best studied by means of noncomputational and nonrepresentational ideas and explanatory schemes involving, e.g., the tools of Dynamical Systems theory.

Versions of this thesis can be found in recent work in developmental psychology (chapter 2 above; Thelen and Smith 1994; Thelen 1995), in work on real-world robotics and autonomous agent theory (chapter 1 above; Smithers 1994; Brooks 1991), in philosophical and cognitive scientific treatments (Maturana and Varela 1987; Varela et al. 1991; Wheeler 1994), and in some neuroscientific approaches (Skarda and Freeman 1987). More circumspect treatments that nonetheless tend toward skepticism about computation and internal representation include Beer and Gallagher 1992, Beer 1995b, van Gelder 1995, and several essays in Port and van Gelder 1995. Historical precedents for such skepticism are also in vogue—see especially Heidegger 1927, Merleau-Ponty 1942, and the works of J. J. Gibson[7] and the ecological psychologists.

The thesis of Radical Embodied Cognition is thus, it seems, a genuinely held view.[8] It involves a rejection of explanations that invoke internal representations, a rejection of computational explanation in psychology, and a suggestion that we would be better off abandoning these old tools in favor of the shining new implements of Dynamical Systems theory.

Such radicalism, I believe, is both unwarranted and somewhat counterproductive. It invites competition where progress demands cooperation. In most cases, at least, the emerging emphasis on the roles of body and world can be seen as complementary to the search for computational and representational understandings. In the next several sections I shall examine a variety of possible motivations for representational and computational skepticism, and I shall show that, in general, the radical conclusion is not justified—either because the phenomena concerned were insufficiently "representation-hungry" in the first place or because the skeptical conclusion depends on too narrow and restrictive a reading of the key terms "representation" and "computation."

8.3 Action-Oriented Representation

One thing is increasingly clear. To the extent that the biological brain does trade in anything usefully described as "internal representation," a large body of those representations will be *local* and *action-oriented* rather than objective and action-independent. One contrast here is between computationally cheap, locally effective "personalized" representations and structures whose content is closer to those of classical symbols signifying elements of public, objective reality. Take the animate-vision work introduced in chapter 1. Here the commitment to the use of both computational and representational descriptions is maintained, but the nature of the computations and representations is reconceived so as to reflect the profound role of (e.g.) actual bodily motion (including foveation) in shaping and simplifying the information-processing problems to be solved. The strategies pursued involved greater reliance on what Ballard (1991) has termed "personalized representations," viz., the use of representations of idiosyncratic, locally effective features to guide behavior. For example, to guide a visual search for your coffee cup you may rely heavily on the particular color of the cup, and you may (according to animate-vision theory) use an internal representation whose content is (e.g.) that *my* cup is yellow. Such a description is only locally effective (it won't generalize to help you find other coffee cups) and is heavily agent-o-centric. But, on the plus side, it involves features that are computationally cheap to detect—color can be spotted even at the low-resolution peripheries of the visual field.

Classical systems, Ballard points out, tended to ignore such locally effective features because the object's identity (its being a coffee cup) does not depend on, e.g., its color. Such systems focused instead on identification procedures which searched for less accidental features and which hence invoked internal representational states whose contents reflected deeper, more agent-independent properties, such as shape and capacity. This emphasis, however, may be inappropriate in any model of the knowledge we actually use on line[9] to guide our real-time searches. In particular, the classical emphasis neglects the pervasive tendency of human agents to *actively structure* their environments in ways that will reduce subsequent computational loads. Thus, it is plausible to suppose that some of us use brightly colored coffee mugs partly *because* this enables us to rely on simple, personalized representations to guide search and identification. In so doing, we are adding structure to our environment in a way designed to simplify subsequent problem-solving behavior—much as certain social insects use chemical trails to add easily usable structure to their local environments, so they can make the route to the food detectable with minimal computational efforts. The commercial sector, as Ballard is quick to point out, displays a profound appreciation (bordering on reverence) for the power of such "nonessential" structurings—Kodak film comes in off-yellow boxes; ecologically sound products display soft green colors on the outside; Brand X corn flakes come in red cartons. Many of these codings will not generalize beyond one supermarket chain or one brand, yet they are used by the consumer to simplify search and identification, and thus they benefit the manufacturer in concrete financial terms.

An animate-vision-style account of the computations underlying everyday on-line visual search and identification may thus fall short of invoking internal representations that describe interpersonally valid defining features of classes of object. Nonetheless, these accounts still invoke stored databases in which specific kinds of locally effective features are associated with target items. Moreover, as we saw in chapter 7, a full account of visual processing will also have to include stories about general mechanisms for the detection of (e.g.) shape, color, and motion. And such stories, we saw, typically associate specific neural pathways and sites with the processing and encoding of specific types of information.

It is also important to be clear, at this point, that the computational and representational commitments of animate-vision research fall well short of requiring us to accept an image of the brain as a classical rule-and-symbol system. Connectionist approaches (chapter 3 above) constitute a clear existence proof of the possibility of alternative frameworks which are nevertheless both computational and (usually) representational. It should also be clear that the kinds of neuroscientific story scouted in chapter 7 are a far cry from the classical vision of a centralized, symbol-crunching inner economy. Witness the depiction in the Convergence Zone hypothesis of higher representational functions as essentially emergent from the time-locked coactivation of more basic, low-level processing regions; or the recognition in the story about primate vision of the context sensitivity, complexity, and sophistication of natural encoding schemes and the accompanying image of neurons as filters tuned along multiple stimulus dimensions; or the role in the new model of monkey finger control of highly distributed and basic-action-oriented forms of internal representation (viz., the representation of whole-hand synergies as the basic subject matter of encodings in the M1 hand area). In all these cases, we confront a neuroscientific vision that is representational (in that it recognizes the need to assign specific content-bearing roles to inner components and processes) *and* that displays a laudable liberality concerning the often complex and non-intuitive ways in which nature deploys its inner resources.

A related contrast, introduced in some detail in previous chapters, concerns the use of representations that are not personalized or partial so much as *action-oriented*. These are internal states which (as described in section 2.6) are simultaneously encodings of how the world is and specifications for appropriate classes of action. The internal map that is already a specification of the motor activity needed to link different locations (section 2.6) is a case in point.

As a further example, consider briefly the robot crab described in chapter 5 of Churchland 1989 and discussed in Hooker et al. 1992. The crab uses point-to-point linkages between two deformed topographic maps so as to directly specify reaching behaviors on the basis of simple perceptual input. In this example, the early visual encodings are themselves deformed or skewed so as to reduce the complexity of the computations required to use that information to specify reaching. This is again to exploit a kind of action-centered (or "deictic"—see Agre 1988) internal representation in which the system does not first create a full, objective world model and then define a costly procedure that (e.g.) takes the model as input and generates food-seeking actions as output. Instead, the system's early encodings are already geared toward the production of appropriate action. This type of action-oriented bias may be at least part of what Gibson was getting at with the rhetorically problematic talk of organisms as "directly perceiving" the world in terms of its affordances for action.[10] Perception, it seems, should not (or, at least, should not always) be conceptualized independently of thinking about the class of *actions* which the creature needs to perform.

Action-oriented representations thus exhibit both benefits and costs. The benefits, as we have noted, involve the capacity to support the computationally cheap guidance of appropriate actions in ecologically normal circumstances. The costs are equally obvious. If a creature needs to use the same body of information to drive multiple or open-ended types of activity, it will often be economical to deploy a more action-neutral encoding which can then act as input to a whole variety of more specific computational routines. For example, if knowledge about an object's location is to be used for a multitude of different purposes, it may be most efficient to generate a single, action-independent inner map that can be accessed by multiple, more special-purpose routines.

It is, however, reasonable to suppose that the more action-oriented species of internal representation are at the very least the most evolutionary and developmentally basic kinds.[11] And it may even be the case that the vast majority of fast, fluent daily problem solving and action depends on them. The point, for present purposes, is just that, even if this turned out to be true, it would fall well short of establishing the thesis of radical embodied cognition, for we would still gain considerable explanatory leverage by understanding the specific features of personal or egocentric space implicated in the action-based analysis. Understanding the specific contents of the action-oriented representations thus plays the usual explanatory role of revealing the adaptive function of certain inner states or processes and helping to fix their contribution to ever-larger information-processing webs.

8.4 Programs, Forces, and Partial Programs

Consider next some of Thelen and Smith's claims (see chapter 2 above) concerning a sense in which learning to walk and learning to reach do not depend on stored programs. Recall that Thelen and Smith showed, rather convincingly, that learning to walk and to reach both depend on a multiplicity of factors spread out across brain, body, and local environment. The picture they present does indeed differ deeply from more traditional views in which the various stages of (e.g.) walking are depicted as the mere playing out of a set of prior instructions encoded in some genetically specified inner resource. The difference lies in the way the child's behavior patterns[12] are seen *not* as under the control of a fixed inner resource but rather as emergent out of "a continual dialogue" involving neural, bodily, and environmental factors. For example, we saw how infant stepping can be induced outside its usual developmental window if the infant is held upright in warm water. And we saw how differences in the energy levels and basic arm-motion repertoires of different infants cause them to confront rather different problems in learning to reach. (The motorically active child must learn to dampen and control its arm flapping, whereas the more passive child must learn to generate sufficient initial force to propel the arm toward a target.) A close study of individual infants exhibiting such differing patterns of motor activity was said to support the general conclusion that solutions to the reaching problem "were discovered in relation to [the children's] own situations, carved out of their individual landscapes, and not prefigured by a synergy known ahead by the brain or the genes" (Thelen and Smith 1994, p. 260). This, of course, is not to say that the solutions had nothing in common. What they have in common, Thelen and Smith plausibly suggest, is a learning routine in which the arm-and-muscle system is treated like an assembly of springs and masses, and in which the job of the central nervous system is to learn to control that assembly by adjusting parameters such as the initial stiffness of the springs. It is thus to coopt the intrinsic dynamics of the system in the service of some current goal. There are thus "no explicit *a priori* instructions or programs for either the trajectory of the hand, joint-angle coordinations, or muscle firing patterns" (ibid., p. 264). Instead, we are said to learn to manipulate a few basic parameters (such as the initial stiffness conditions) so as to sculpt and modulate the behavior of a changing physical system with rich and developing intrinsic dynamics.

The experiments and data displayed by Thelen and Smith are fascinating, important, and compelling. But they do not unambiguously support the complex of radical claims I outlined in section 8.2. Rather than amounting to a clear case against computationalism and representationalism *in general*, what we confront is another body of evidence suggesting that we will not discover the *right* computational and representational stories unless we give due weight to the role of body and local environment—a role that includes both problem definition and, on occasion, problem solution. We saw how the spring-like qualities of infant muscles and the varying levels of infant energy help fix the specific problems a given brain must solve. And it is easy to imagine ways in which bodily and environmental parameters may likewise contribute to specific solutions—for example, compliance in spring-mounted limbs allows walking robots to adjust to uneven terrains without the massive computational effort that would be needed to achieve the same result using sensors and feedback loops in a noncompliant medium (see also Michie and Johnson 1984 and chapter 1 above). Moreover, once we recognize the role of body and environment (recall the examples of stepping infants suspended in water) in constructing both problems and solutions, it becomes clear that, for certain explanatory purposes, the *overall* system of brain, body, and local environment can constitute a proper, unified object of study. Nonetheless, this whole complex of important insights is fully compatible with a computational[13] and representational approach to the study of cognition. The real upshot of these kinds of consideration is, I suggest, a better kind of computational and representational story, not the outright rejection of computationalism and representationalism.

It is revealing, with this in mind, to look more closely at some of the specific passages in which computationalism is rejected. A typical example follows:

> The developmental data are compelling in support of . . . anticomputational views. What is required is to reject both Piaget's vision of the end-state of development as looking like a Swiss logician, and the maturationist conviction that there is an executive in the brain . . . that directs the course of development. (Thelen 1995, p. 76)[14]

The point to notice is that here (as elsewhere) the bold assertion of anti-computationalism is followed by a narrower and more accurate description of the target. If we ignore the sweeping claims and focus on these narrower descriptions, we find that the real villain is not computationalism (or representationalism) *per se* but rather

(1) the claim that development is driven by a fully detailed advance plan

and

(2) the claim that adult cognition involves internal logical operations on propositional data structures (i.e., the logicist view of the endpoint of cognitive development attributed to Piaget).

In place of these theses, Thelen and Smith suggest (quite plausibly, in my view) the following:

(1*) Development (and action) exhibit order which is merely executory. Solutions are "soft assembled" out of multiple heterogeneous components including bodily mechanics, neural states and processes, and environmental conditions. (1994, p. 311)

(2*) Even where adult cognition *looks* highly logical and propositional, it is actually relying on resources (such as metaphors of force, action, and motion) developed in real-time activity and based on bodily experience. (ibid., p. 323; Thelen 1995)

Experimental support for (1) was adduced in chapter 2 above. I shall not rehearse the case for (2), nor shall I attempt to further adjudicate either claim. Instead, I note only that granting both claims is quite compatible with a substantial commitment to the use of computational explanation. In fact, the evidence presented works best against models involving classical quasi-linguistic encodings and against the view that initial states of the child's head fully determine the course of subsequent development. Yet good old-fashioned connectionism has long disputed the tyranny of quasi-linguistic encodings. And the computationally inclined developmentalist can surely still embrace an idea (to be cashed out below) of *partially programmed* solutions—that is, cases where the child's initial program is set up by evolution precisely so as to allow bodily dynamics and local environmental contingencies to help determine the course and the outcome of the developmental process. Partial programs would thus share the logical character of most genes: they would fall short of constituting a full blueprint of the final product, and would cede many decisions to local environmental conditions and processes. Nonetheless, they would continue to constitute isolable factors which, in a natural setting, often make a "typical and important difference."[15]

Thus, suppose we concede that, at least in the cases discussed, the brain does not house any comprehensive recipes for behavioral success. Would it follow that talk of internal programs is necessarily misguided in such cases? And even if it did, would it also follow that the neural roots of such activity are not usefully seen as involving computational processes? I shall argue that the answer to both questions is No. In short, I question both the implicit transition from "no comprehensive recipe" to "no inner program" and that from "no inner program" to "no computation." These moves strike me as deeply flawed, although they raises a constellation of deep and subtle issues which I do not pretend to be able to fully resolve.

Consider the very idea of a *program* for doing such and such—for example, calculating your tax liability. The most basic image here is the image of a recipe—a set of instructions which, if faithfully followed, will solve the problem. What is the difference between a recipe and a *force* which, if applied, has a certain result? Take, for example, the heat applied to a pan of oil: the heat will, at some critical value, cause the emergence of swirls, eddies, and convection rolls in the oil. Is the heat (at critical value) a program for the creation of these effects? Is it a recipe for swirls, eddies, and convection rolls? Surely not—it is just a force applied to a physical system. The contrast is obvious, yet it is surprisingly hard to give a principled account of the difference. Where should we look to find the differences that make the difference?

One place to turn is to the idea of a program as, literally, a set of instructions. Instructions are couched in some kind of a language—a system of signs that can be interpreted by some kind of reading device (a hearer, for instructions in English; a compiler, for instructions in LISP; and so on). One reason why the heat applied to the pan does not seem like a program for convection rolls may thus be that we see no evidence of a language here—no evidence of any signs or signals in need of subsequent interpretation or decoding.

Another, not unrelated, reason is that the guiding parameter (the amount of heat needed to produce, e.g., convection rolls) seems too simple and undifferentiated. It is, as one of my students usefully remarked, more like plugging the computer in than running a program on it. Indeed, it is my suspicion (but here is where the waters are still too murky to see clearly) that this difference is the fundamental one, and that the point about requiring a language or code is somehow derivative. Consider two putative "programs" for calculating your tax liability. One consists of 400 lines of code and explicitly covers all the ground you would expect. The other is designed to be run on a very special piece of hardware which is already set up to compute tax functions. This "program" consists solely of the command "Do Tax!" This is surely at best a marginal or limiting case of a program. It is an instruction that (let's assume) needs to be decoded by a reading device before the desired behavior ensues. But it seems to have more in common with the "mere plugging in" scenario (and hence with the heat-to-a-pan model) than with the image of a recipe for success. Perhaps, then, the conceptual bedrock is not about the mere involvement of signs and decodings but rather about the extent to which the target behavior (the tax calculation, the convection rolls) is actually specified by the applied force rather than merely prompted by it. Such a diagnosis seems intuitively appealing and would help explain why, for example, it is at least tempting to treat DNA as programming physical outcomes,[16] while denying that the heat programs the oil in the pan.

The idea of a *partial* program is thus the idea of a genuine specification that nonetheless cedes a good deal of work and decision making to other parts of the overall causal matrix. In this sense, it is much like a regular computer program (written in, say, LISP) that does not specify how or when to achieve certain subgoals but instead cedes those tasks to built-in features of the operating system. (Indeed, it is only fair to note that no computer program ever provides a full specification of how to solve a problem—at some point or points, specification gives out and things just happen in ways determined by the operating system or the hardware.) The phrase "partial program" thus serves mainly to mark out the rather special class of cases in which some such decisions and procedures are ceded to rather more distant structures: structures in the wider causal matrix of body and external world. For example, a motor control system such as the emulator circuitry described in Jordan et al. 1994 and discussed in chapter 1 above may be properly said to learn a program for the control of arm trajectories. Yet it is a program that will yield success only if there is a specific backdrop of bodily dynamics (mass of arm, spring of muscles) and environmental features (force of gravity). It is usefully seen as a program to the extent that it nonetheless specifies reaching motions in a kind of neural vocabulary. The less detailed the specification required (the more work is being done by the intrinsic—long-term or temporary—dynamics of the system), the less we need treat it as a program. We thus confront not a dichotomy between programmed and unprogrammed solutions so much as a continuum in which solutions can be more or less programmed according to the degree to which some desired result depends on a series of moves (either logical or physical) that require actual specification rather than mere prompting.

In pursuing this contrast, however, we must bear in mind the very real possibility of a kind of cascade of computational activity in which a simple unstructured command is progressively unpacked, via a sequence of subordinate systems, into a highly detailed specification that ultimately controls behavior (see, e.g., Greene 1972 and Gallistel 1980). Should such progressive unpacking occur in the chain of neural events, we may count the stage (or stages) of more detailed specification as the stored program. The controversial claim of Thelen and Smith, Kelso, and others is that even the most detailed stages of neural specification may not be worth treating as a stored program—that so much is done by the synergetic[17] dynamics of the bodily system that the neural commands are at all stages best understood as the application of simple forces to a complex body-environment system whose own dynamics carry much of the problem-solving load. Less radically, however, it may be that what these investigations really demonstrate is that the problem of bringing about reaching motions and the like may require somewhat less in the way of detailed inner instruction sets than we hitherto supposed, courtesy of the rather complex synergetic dynamics already implemented in (e.g.) the arms and muscles themselves. Perhaps, as Thelen and Smith claim, the kind of specification required to generate some motor behaviors amounts only to specifying the settings of a few central parameters (initial stiffness in the spring-like muscle system, etc.); these sparse kinds of specification may then have complex effects on the overall dynamics of the physical system such that reaching can be achieved without directly specifying things like joint-angle configurations. The point to notice is that the lack of a particular *kind* of specification or instruction set (e.g., the kind that explicitly dictates joint-angle configurations and muscle-firing patterns) does not itself establish the complete lack of any specification or program. Indeed, such a characterization looks most compelling only at the extreme limiting case in which the notion of a coded specification collapses into the notion of a simple applied force or a single unstructured command. There is thus plenty of very interesting space to explore between the idea of a stored program that specifies a problem-solving strategy at a very low level (e.g., the level of muscle firing patterns) and the idea of a system whose intrinsic dynamics render specification altogether unnecessary or (what really amounts to the same thing) reduce it to the level of applying a simple force. Between these two extremes lies the space of what I have called "partial programs." The real moral of dynamic-systems-oriented work in motor control, I believe, is that this is the space in which we will find nature's own programs.

But suppose we don't. Suppose there exists no level of neural elaboration of commands worth designating as a "stored program." Even then, I suggest, it would not follow that the image of the brain as a *computing device* is unsound. It is, alas, one of the scandals of cognitive science that after all these years the very idea of computation remains poorly understood. Given this lack of clarity, it is impossible to make a watertight case here. But one attractive option is to embrace a notion of computation that is closely tied to the idea of automated information processing and the mechanistic transformation of representations. According to this kind of account, we would find computation whenever we found a mechanistically governed transition between representations, irrespective of whether those representations participate in a specification scheme that is sufficiently detailed to count as a stored program. In addition, this relatively liberal[18] notion of computation allows easily for a variety of styles of computation spanning both digital computation (defined over discrete states) and analog computation (defined over continuous quantities). On this account, the burden of showing that a system is computational reduces to the task of showing that it is engaged in the automated processing and transformation of information.

In conclusion, it seems multiply premature to move from the kinds of image and evidence adduced by Thelen and Smith and others to the conclusion that we should abandon notions of internal representation and computation in our efforts to understand biological cognition. Instead, what really emerges from this work and from the work in animate vision and robotics discussed earlier is a pair of now-familiar but very important cautions, which may be summed up as follows:

(1) Beware of putting too much into the head (or the inner representational system). What gets internally represented and/or computed over will be determined by a complex balancing act that coopts both bodily and environmental factors into the problem-solving routine. As a result, both partially programmed solutions and action-oriented or personalized representations will be the order of the biological day.
(2) Beware of rigid assumptions concerning the form of internal representations or the style of neural computation. There is no reason to suppose that classical (spatio-temporally localized) representations and discrete, serial computation limn the space of representational and computational solutions. Connectionist models have, in any case, already begun to relax these constraints—and they merely scratch the surface of the range of possibilities open to biological systems.

8.5 Beating Time

Perhaps there are other reasons to be wary of representational approaches. Such approaches, it has recently been argued, cannot do justice to the crucial temporal dimensions of real-world adaptive response. (See especially the introduction to Port and van Gelder 1995). Early connectionist models, for example (see chapter 3), displayed no intrinsic knowledge of time or order and depended on a variety of tricks[19] to disambiguate sequences containing identical elements. Moreover, the inputs to such networks were instantaneous "snapshots" of the world, and the outputs were never essentially temporally extended patterns of activity. The advent of *recurrent* networks (Jordan 1986; Elman 1991) signaled a degree of progress, since such networks incorporated internal feedback loops which enabled responses to new inputs to take account of the networks previous activities. Yet, as Port et al. (1995) nicely point out, such networks are good at dealing with order rather than with real timing. For example, such networks are able to specify an ordered sequence of operations as outputs (e.g. outputting a sequence of instructions to draw a rectangle—see Jordan 1986) or to display a sensitivity to grammatical constraints that depend on the order of inputs (see Elman's 1991 work on artificial grammars). But order is not the same thing as real timing. In running to catch a moving bus, you must do more than produce the right sequence of motor commands. You must register a pattern unfolding over time (the bus accelerating away from you) and generate a range of compensating actions (a temporally coordinated sequence of motor commands to legs, arms, and body). And at the point of contact (if you are so lucky) there must be a delicate coupling between the temporally extended activity of the two systems (you and the bus). Such behavior requires at least one system to respond to the *real timing* (not just the order) of events in the other. To model this, researchers have begun to find ways to use the real-time properties of incoming signals to "set" inner resources. The trick is to use the real timing of some input signals as a "clock" against which to measure other such signals. One way this can be achieved is by developing an "adaptive oscillator." Such devices (Torras 1985; McCauley 1994; Port et al. 1995) have two key properties. First, they generate periodic outputs all on their own (like neurons that have a tonic spiking frequency). Second, this periodic activity can be affected by incoming signals. Should such an oscillator detect incoming signals, it fires (spikes) immediately and alters its periodicity to bring it slightly more in line with that of the incoming signals. Within fixed boundaries such an oscillator will, over time, come to fire perfectly in phase with the inputs. Should the input signal cease, the oscillator will gradually "let go," returning to its natural rate. Neural network versions of such devices adapt by using the familiar gradient descent learning procedure. But the information that powers the descent is in this case the difference between the usual ("expected") timing of a spike and the real timing caused by the device's tendency to fire at once if an input signal is detected. Such devices thus "entrain" to the frequency of a detected signal and can then maintain that frequency for a while even if the signal disappears or misses a beat. Entrainment is not immediate, so nonperiodic signals have no real effects (they produce one unusually timed spike and that's it). But regular signals cause the device to "pick up the rhythm." A complex system may deploy many adaptive oscillators, each having a different natural tempo and hence being especially sensitive to different incoming signal ratios. Global entrainment to a stimulus that contains several different periodic elements (e.g. a piece of music) occurs when several individual oscillators latch onto different elements of the temporal structure.[20]

For present purposes, the lesson of all this is that internal processes, with intrinsic temporal features, may figure prominently in the explanation of an important subset of adaptive behaviors. In such cases, the "fit" between the inner state and the external circumstances may indeed go beyond anything captured in the usual notion of internal representation. The adaptive oscillator does its job by coupling its activity to the rhythms of external events yielding periodic signals. It does not represent that periodicity by the use of any arbitrary symbol, still less by the use of any text-like encoding. Instead, it is best seen as an internal system suited to temporarily merging with external systems by parasitizing their real temporal properties. In attempting to analyze and explain such capacities, we need both the perspective in which the external system is a source of inputs to entrain the oscillator and the perspective that focuses on the subsequent properties of the larger, coupled system. Yet, despite these complications, it surely remains both natural and informative to depict the oscillator as a device whose adaptive role is to represent the temporal dynamics of some external system or of specific external events. The temporal features of external processes and events are, after all, every bit as real as colors, weights, orientations, and all the more familiar targets of neural encodings. It is, nonetheless, especially clear in this case that the kind of representation involved differs from standard conceptions: the vehicle of representation is a *process*, with *intrinsic temporal* properties. It is not an arbitrary vector or symbol structure, and it does not form part of a quasi-linguistic system of encodings. Perhaps these differences are enough to persuade some theorists that such processes are not properly termed *representational* at all. And it is not, ultimately worth fighting over the word. It does seem clear, however, that we understand the role and function of the oscillator only by understanding what kinds of features of external events and processes it is keyed to and, hence, what other neural systems and motor-control systems might achieve by consuming the information it carries.

8.6 Continuous Reciprocal Causation

There is one last way (that I know of) to try to make the strong anti-representationalist case. It involves an appeal to to the presence of continuous, mutually modulatory influences linking brain, body, and world. We have already encountered hints of such mutually modulatory complexity in the interior workings of the brain itself (see the discussion of mammalian vision in section 7.3). But suppose something like this level of interactive complexity characterized some of the links among neural circuitry, physical bodies, and aspects of the local environment? The class of cases I have in mind can be gently introduced by adapting an analogy due to Tim van Gelder (personal communication).

Consider a radio receiver, the input signal to which is best treated as a continuous modulator of the radio's "behavior" (its sound output). Now imagine (here is where I adapt the analogy to press the point) that the radio's output is also a continuous modulator of the external device (the transmitter) delivering the input signal. In such a case, we observe a truly complex and temporally dense interplay between the two system components—one which could lead to different overall dynamics (e.g. of positive feedback or stable equilibria) depending on the precise details of the interplay. The key fact is that, given the continuous nature of the mutual modulations, a common analytic strategy yields scant rewards. The common strategy is, of course, componential analysis, as described in chapter 6. To be sure, we can and should identify different components here. But the strategy breaks down if we then try to understand the behavioral unfolding of one favored component (say, the receiver) by treating it as a unit *insulated*[21] from its local environment by the traditional boundaries of transduction and action, for such boundaries, in view of the facts of continuous mutual modulation, look arbitrary with respect to this specific behavioral unfolding. They would not be arbitrary if, for example, the receiver unit displayed discrete time-stepped behaviors of signal receiving and subsequent broadcast. Were that the case, we could reconceptualize the surrounding events as the world's giving inputs to a device which then gives outputs ("actions") which affect the world and hence help mold the next input down the line—for example, we could develop an interactive "catch and toss" version of the componential analysis, as predicted in chapter 6.

A second example (this one was suggested to me by Randy Beer) may help fix the difference. Consider a simple two-neuron system. Suppose that neither neuron, in isolation, exhibits any tendency toward rhythmic oscillation. Nonetheless, it is sometimes the case that two such neurons, when linked by some process of continuous signaling, will modulate each other's behavior so as to yield oscillatory dynamics. Call neuron 1 "the brain" and neuron 2 "the environment." What concrete value would such a division have for understanding the oscillatory behavior?

Certainly there are two components here, and it is useful to distinguish them and even to study their individual dynamics. However, for the purpose of explaining the oscillation there is nothing special about the "brain" neuron. We could just as well choose to treat the other component (the "environment" neuron) as the base-line system and depict the "brain" neuron as merely a source of perturbations to the "environment." The fact of the matter, in this admittedly simplistic case, is that neither component enjoys any special status given the project of explaining the rhythmic oscillations. The target property, in this case, really is best understood and studied as an emergent property of the larger system created by the coupling of the two neurons. Similarly, in the case of biological brains and local environments it would indeed be perverse—as Butler (to appear) rightly insists—to pretend that we do not confront distinct *components*. The question, however, must be whether certain target phenomena are best explained by granting a kind of special status to one component (the brain) and treating the other as merely a source of inputs and a space for outputs. In cases where the target behavior involves continuous reciprocal causation between the components, such a strategy seems ill motivated. In such cases, we do not, I concede, confront a single undifferentiated system. But the target phenomenon is an emergent property of the coupling of the two (perfectly real) components, and should not be "assigned" to either alone.

Nor, it seems to me, is continuous reciprocal causation[22] a rare or exceptional case in human problem solving. The players in a jazz trio, when improvising, are immersed in just such a web of causal complexity. Each member's playing is continually responsive to the others' and at the same time exerts its own modulatory force. Dancing, playing interactive sports, and even having a group conversation all sometimes exhibit the kind of mutually modulatory dynamics which look to reward a wider perspective than one that focuses on one component and treats all the rest as mere inputs and outputs. Of course, these are all cases in which what counts is something like the social environment. But dense reciprocal interactions can equally well characterize our dealings with complex machinery (such as cars and airplanes) or even the ongoing interplay between musician and instrument. What matters is not whether the other component is itself a cognitive system but the nature of the causal coupling between components. Where that coupling provides for continuous and mutually modularity exchange, it will often be fruitful to consider the emergent dynamics of the overarching system.

Thus, to the extent that brain, body, and world can at times be joint participants in episodes of dense reciprocal causal influence, we will confront behavioral unfoldings that resist explanation in terms of inputs to and outputs from a supposedly insulated individual cognitive engine. What would this mean for the use in such cases of the notion of internal representation in cognitive scientific explanations? There would seem to be just two possibilities.

The first is that we might nonetheless manage to motivate a representational gloss for some specific subset of the agent-side structures involved. Imagine a complex neural network, A, whose environmentally coupled dynamics include a specific spiking (firing) frequency which is used by other onboard networks as a source of information concerning the presence or absence of certain external environmental processes—the ones with which A is so closely coupled. The downstream networks thus use the response profiles of A as a stand-in for these environmental states of affairs. Imagine also that the coupled response profiles of A can sometimes be induced, in the absence of the environmental inputs, by top-down neural influences,[23] and that when this happens the agent finds herself imagining engaging in the complex interaction in question (e.g., playing in a jazz trio). In such circumstances, it seems natural and informative to treat A as a locus of internal representations, despite its involvement, at times, in episodes of dense reciprocal interaction with external events and processes.

A second possibility, however, is that the system simply never exhibits the kind of potentially decoupled inner evolution just described. This will be the case if, for example, certain inner resources participate *only* in densely coupled, continuous reciprocal environmental exchanges, and there seem to be no identifiable inner states or processes whose role in those interactions is to carry specific items of information about the outer events. Instead, the inner and the outer interact in adaptively valuable ways which simply fail to succumb to our attempts to fix determinate information-processing roles to specific purely internal, components, states, or processes. In such a case the system displays what might be called nonrepresentational adaptive equilibrium. (A homely example is a tug of war: neither team is usefully thought of as a representation of the force being exerted by the other side, yet until the final collapse the two sets of forces influence and maintain each other in a very finely balanced way.)

Where the inner and the outer exhibit this kind of continuous, mutually modulatory, non-decouplable coevolution, the tools of information-processing decomposition are, I believe, at their weakest. What matters in such cases are the real, temporally rich properties of the ongoing exchange between organism and environment. Such cases are conceptually very interesting, but they do not constitute a serious challenge to the general role of representation-based understanding in cognitive science. Indeed, they cannot constitute such a challenge, since they lie, by definition, outside the class of cases for which a representational approach is most strongly indicated, as we shall now see.

8.7 Representation-Hungry Problems

The most potent challenge to a representation-based understanding comes, we saw, from cases in which the web of causal influence grows so wide and complex that it becomes practically impossible to isolate any "privileged elements" on which to pin specific information-carrying adaptive roles. Such cases typically involve the continuous, reciprocal evolution of multiple tightly linked systems, whose cumulative ('emergent') effect is to promote some kind of useful behavior or response. In seeking to do justice to such problematic cases, however, we should not forget the equally compelling range of cases for which a representational understanding seems *most* appropriate. These are the cases involving what I elsewhere[24] dub "representation-hungry problems."

Recall the first of Haugeland's strong requirements for an internal-representation-using system (section 8.2 above). It was that the system must coordinate its behaviors with environmental features that are not always "reliably present to the system." There are, I think, two main classes of cases in which this constraint is met. These are (1) cases that involve reasoning about absent, nonexistent, or counterfactual states of affairs and/or (2) cases that involve selective sensitivity to states of affairs whose physical manifestations are complex and unruly.

The first class of cases (already mentioned in section 8.2) include thoughts about temporally or spatially distant events and thoughts about the potential outcomes of imagined actions. In such cases, it is hard to avoid the conclusion that successful reasoning involves creating some kind of prior and identifiable stand-ins for the absent phenomena—inner surrogates that make possible appropriate behavioral coordination without the guidance provided by constant external input.

The second class of cases (which Haugeland does not consider) is equally familiar, although a little harder to describe. These are cases in which the cognitive system must selectively respond to states of affairs whose physical manifestations are wildly various—states of affairs that are unified at some rather abstract level, but whose physical correlates have little in common. Examples would include the ability to pick out all the valuable items in a room and the ability to reason about all and only the goods belonging to the pope. It is very hard to see how to get a system to reason about such things without setting it up so that all the various superficially different inputs are first assimilated to a common inner state or process such that further processing (reasoning) can then be defined over the inner correlate: an inner item, pattern, or process whose content then corresponds to the abstract property. In such cases, behavioral success appears to depend on our ability to compress or dilate a sensory input space. The successful agent must learn to treat inputs whose early encodings (at the sensory peripheries) are very different as calling for the same classification, or, conversely, to treat inputs whose early encodings are very similar as calling for different classifications. Identifiable internal states developed to serve such ends just are internal representations whose contents concern the target (elusive) states of affairs in question.[25] (Should any such story prove correct, it would be hard to resist the conclusion that even basic visual recognition involves, at times, computations defined over genuine internal representational states).

In the two ranges of cases (the absent and the unruly), the common feature is the need to generate an additional internal state whose information-processing adaptive role is to guide behavior despite the effective unfriendliness of the ambient environmental signals (either there are none, or they require significant computation to yield useful guides for action). In these representation-hungry cases, the system must, it seems, create some kind of inner item, pattern, or process whose role is to stand in for the elusive state of affairs. These, then, are the cases in which it is most natural to expect to find system states that count as full-blooded internal representations.

It may seem, indeed, that in such cases there cannot fail to be internal representations underlying behavioral success. This, however, is too strong a conclusion, for there is surely an important pragmatic element that might still confound the attempt to make sense of the system in representational terms. Thus, although the representation-hungry cases clearly demand that some kind of systemic property compensate for the lack of reliable or easily usable ambient inputs, it does not follow that the relevant property will be *usefully individuable*. It will not be usefully individuable if, once again, it somehow involves such temporally complex and reciprocally influential activity across so many subsystems that the "standing in" is best seen as an emergent property of the overall operation of the system. In such cases (if there are any), the overall system would rightly be said to represent its world—but it would not do so by trading in anything we could usefully treat as internal representations. The notion of internal representation thus gets a grip only when we can make relatively fine-grained assignments of inner vehicles to information-carrying adaptive roles. Such vehicles may be spatially distributed (as in the hypothesis of convergence zones), they may be temporally complex, and they may involve analog qualities and continuous numerical values. But they must be identifiable as distinct subsets of overall systemic structure or activity. My guess (consistent with the state of contemporary neuroscience) is that such identification will prove possible, and that it will play a crucial role in helping us to understand certain aspects of our adaptive success. At the very least, we can now see more clearly what it would take to undermine a representation-based approach: it would require a demonstration that, even in the representation-hungry cases, it remains practically impossible to isolate any system of fine-grained vehicles playing specific information-carrying adaptive roles. Moreover, we have seen many ways in which the fundamental insights of an embodied, embedded approach (concerning action-oriented encodings, environment-involving problem solving, and synergetic couplings between multiple elements) are in any case compatible with the use of computational and representational understandings.

In the final analysis, then, the resolution of this debate must turn on future empirical studies. No doubt there is some upper limit[26] on the degree of complexity of inner states and processes such that beyond that limit it becomes simply uninformative and explanatorily idle to describe them as internal representations. But the question as to exactly where this limit lies will probably be resolved only by practical experience. The answer will emerge by trial and error, as experimentalists generate and analyze real dynamical solutions to increasingly complex and superficially "representation-hungry" problems. Such confrontations may lead to a process of mutual accommodation in which the dynamical systems stories are adapted and enriched with computational and representational forms of understanding and analysis and vice versa.[27] Or the sheer complexity of the dynamical patterns and processes involved, and the deep interweaving of inner and outer elements, may convince us that it is fruitless to try to identify any specific aspects of the complex and shifting causal web as signaling the presence of specific environmental features, and hence fruitless to pursue any representational understanding of the structure and operation of the system. The most likely outcome, it seems to me, is not the outright rejection of ideas of computation and representation so much as their partial reconception. Such a reconception is prefigured in many dynamical analyses of more representation-hungry kinds of problem (such as decision making, and planning[28]) and is a natural continuation of research programs in both connectionism and computational neuroscience.

Such a reconception would, however, have implications that woulc go beyond the mere identification of a new range of inner vehicles capable of playing representational roles. On the positive side, the representational vehicles would no longer be constrained to the realm of inner states and processes. By allowing (e.g.) values of collective variables to take on representational significance, the dynamical theorist can allow some of a system's content-bearing states to be intrinsically wide—to depend on states defined only across the larger system comprising the agent and some select chunks of the local environment.[29] On the negative side, to the extent that the representational vehicles are allowed to float higher and higher above the level of basic system variables and parameters,[30] we may witness the partial fracturing of a powerful and familiar explanatory schema. The worry (already familiar from section 6.4 above) is that we thus begin to dislocate the representational description of a system (and, more generally, its information-processing characterization) from the kind of description that would speak directly to the project of actually building or constructing such a system. In contrast, one of the primary virtues of more standard computational models is that they display the way information and representations flow through the system in ways that are constrained to yield effective recipes for generating such behavior in a real physical device. By allowing representational glosses to stick to complex dynamical entities (limit cycles, state-space trajectories, values of collective variables, etc.), the theorist pitches the information-processing story at a very high level of abstraction from details of basic systemic components and variables, and thus severs the links between the representational description and the specific details of inner workings. The best representational stories, it now seems, may be pitched at an even greater remove[31] from the nitty-gritty of physical implementation than was previously imagined.

8.8 Roots

The anti-representationalist and anti-computationalist intuitions discussed in the preceding sections have a variety of antecedents, both recent and

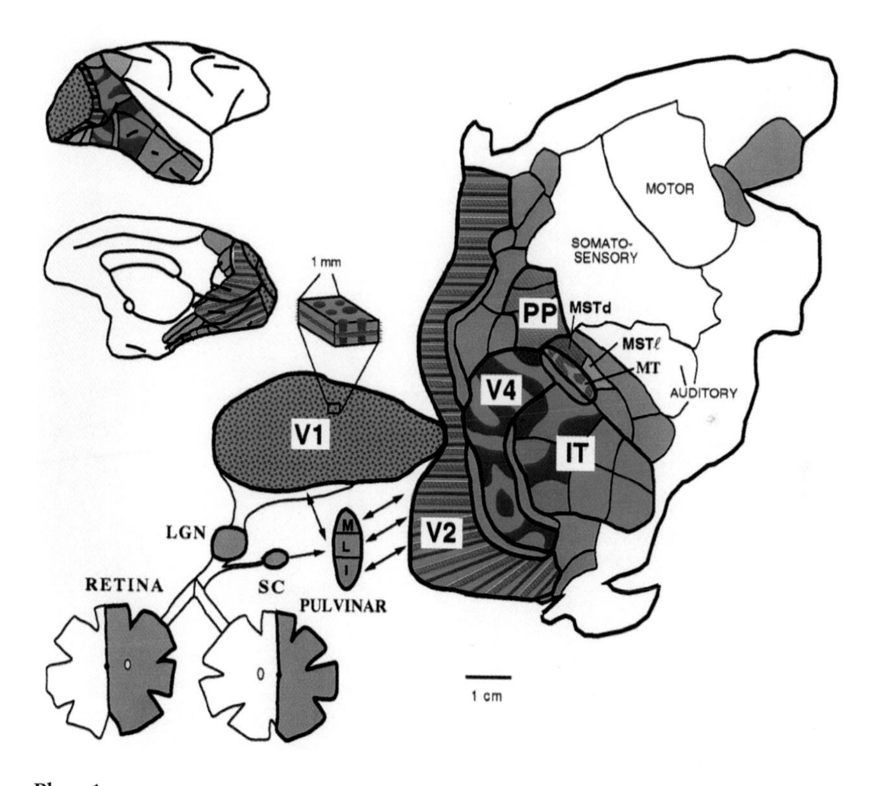

Plate 1

Two-dimensional map of cerebral cortex and major subcortical visual centers of macaque monkey. The flattened cortical map encompasses the entire right hemisphere. Source: Van Essen and Gallant 1994. Courtesy of David Van Essen, Jack Gallant, and Cell Press.

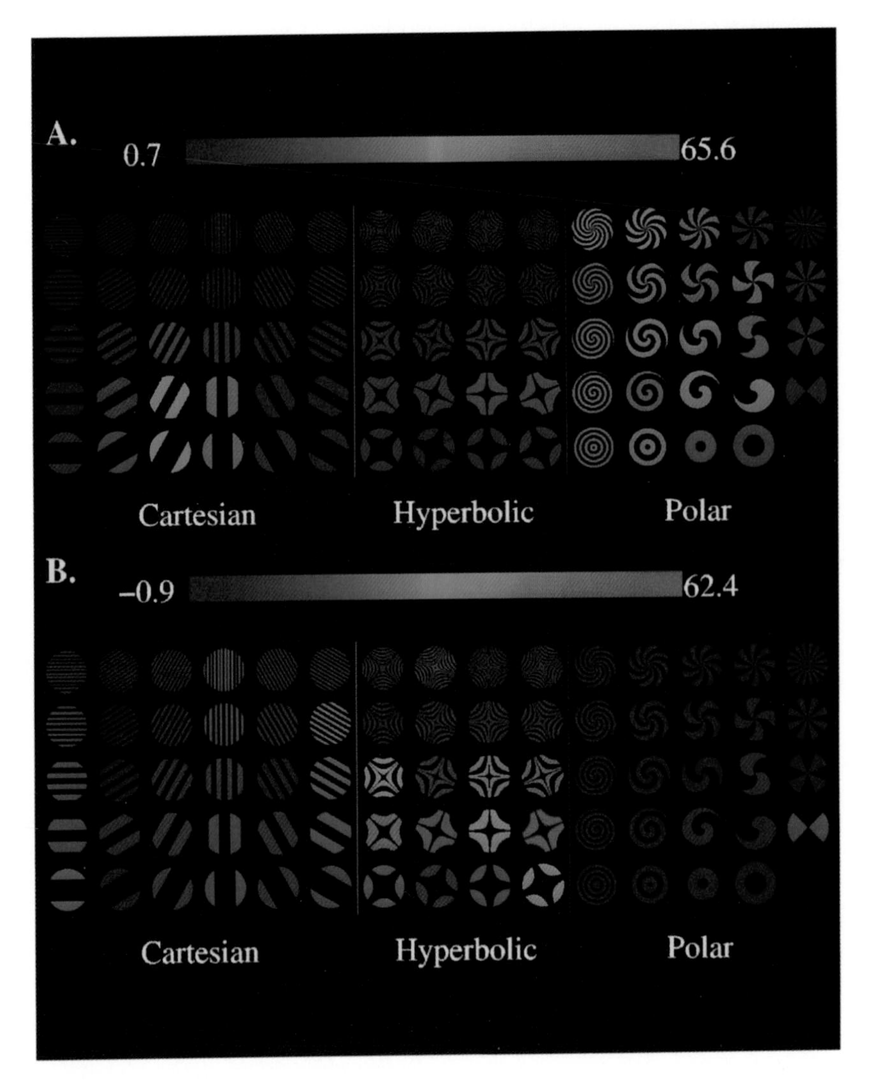

Plate 2
Responses of two cells in V4 to different cartesian and noncartesian spatial patterns. Each icon represents a particular visual stimulus, and its color represents the mean response to that stimulus relative to the spontaneous background rate, using the color scale shown above. (A) A neuron that responded maximally to polar stimuli, particular spirals, and much less hyperbolic and cartesian gratings (modified, with permission, from Gallant et al. 1993). (B) A neuron that responded maximally to hyperbolic stimuli of low to moderate spatial frequencies. Source: Van Essen and Gallant 1994; courtesy of David Van Essen, Jack Gallant, and Cell Press.

Plate 3
Robot tuna hanging from carriage in Ocean Engineering Testing Tank Facility at MIT. Source: Triantafyllou and Triantafyllou 1995. Photograph by Sam Ogden; used by courtesy of Sam Ogden and with permission of Scientific American, Inc.

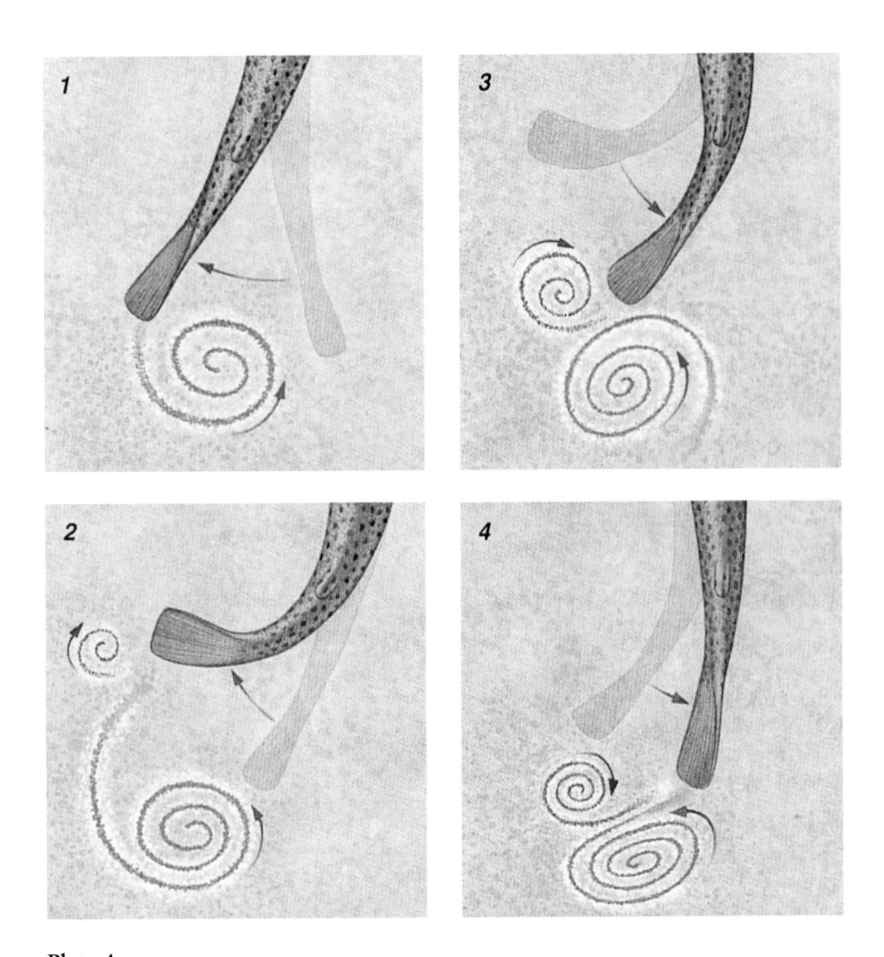

Plate 4

A forceful flap followed in quick succession by another in the reverse direction produces a strong, sudden thrust well suited to pouncing on prey or making a fast getaway. The initial flap makes a large vortex (1); the second flap creates a different, counterrotating vortex (2, 3). A strong forward thrust and a stray but manageable lateral force result when the two vortices meet and combine to create a jet and are pushed away from the tail, weakening each other (4). Source: Triantafyllou and Triantafyllou 1995; courtesy of M. S. and G. S. Triantafyllou and of Scientific American, Inc.

not-so-recent. I will round off the present discussion by sketching some[32] of these roots and displaying some differences of emphasis and scope.

Heidegger (1927) wrote of the importance of *Dasein* (being there)—a mode of being-in-the-world in which we are not detached, passive observers but active participants—and stressed the way our practical dealings with the world (hammering nails, opening doors, and so on) do not involve detached representings (e.g. of the hammer as a rigid object of a certain weight and shape) so much as *functional couplings*. We use the hammer to drive in the nail, and it is this kind of skilled practical engagement with the world that is, for Heidegger, at the heart of all thought and intentionality.[33] A key notion in this analysis is the idea of *equipment*—the stuff that surrounds us and figures in the multiple skilled activities underlying our everyday abilities to cope and succeed.

Thus, Heidegger's work prefigures skepticism concerning what might be termed "action-neutral" kinds of internal representation, and it echoes our emphasis on tool use and on action-oriented couplings between organism and world. Some of Heidegger's concerns, however, are radically different from those of the present treatment. In particular, Heidegger was opposed to the idea that knowledge involves a relation between minds and an independent world (Dreyfus 1991, pp. 48–51)—a somewhat metaphysical question on which I take no stand. In addition, Heidegger's notion of the milieu of embodied action is thoroughly social. My version of being there is significantly broader and includes all cases in which body and local environment appear as elements in extended problem-solving activity.[34]

Closer in spirit and execution to the present project is the work of the phenomenologist Maurice Merleau-Ponty,[35] who was concerned to depict everyday intelligent activity as the playing out of whole organism-body-world synergies. In particular, Merleau-Ponty stressed the importance of what I have called "continuous reciprocal causation"—viz., the idea that we must go beyond the passive image of the organism perceiving the world and recognize the way our actions may be continuously responsive to worldly events which are at the same time being continuously responsive to our actions. Consider a lovely example, which I think of as "the hamster and tongs":

> When my hand follows each effort of a struggling animal while holding an instrument for capturing it, it is clear that each of my movements responds to an external stimulation; but it is also clear that these stimulations could not be received

without the movements by which I expose my receptors to their influence. . . . The properties of the object and the intentions of the subject are not only intermingled; they also constitute a new whole. (Merleau-Ponty 1942, p. 13)

In this example the motions of my hands are continuously responsive to those of the struggling hamster, but the hamster's struggles are continuously molded and shaped by the motions of my hand. Here action and perception, as David Hilditch (1995) has put it, coalesce as a kind of "free form interactive dance between perceiver and perceived." It is this iterated interactive dance that, we saw, is now recognized in recent work concerning the computational foundations of animate vision.

Moreover, Merleau-Ponty also stresses the way perception is geared to the control of real-time, real-world behavior. In this respect, he discovers something very like[36] the Gibsonian notion of an affordance—a notion which, in turn, is the direct inspiration of the idea of action-oriented internal representations discussed above in chapter 2 and in section 8.3. An affordance is an opportunity for use or interaction which some object or state of affairs presents to a certain kind of agent. For example, to a human a chair affords sitting, but to a woodpecker it may afford something quite different.

Gibson's special concern was with the way visual perception might be tuned to invariant features presented in the incoming light signal in ways that directly selected classes of possible actions—for example, the way patterns of light might specify a flat plain affording human walking. To the extent that the human perceptual system might become tuned to such affordances, Gibson claimed that there was no need to invoke internal representations as additional entities mediating between perception and action. In section 8.3 I argued that such outright rejection often flows from an unnecessary conflation of two properly distinct notions. One is the fully general idea of internal representations as inner states, structures, or processes whose adaptive role is to carry specific types of information for use by other neural and action-guiding systems. The other is the more specific idea of internal representations as rich, *action-neutral* encodings of external states of affairs. Only in the latter, more restricted sense is there any conflict between Gibsonian ideas and the theoretical construct of internal representation.[37]

Finally, the recent discussion of "the embodied mind" offered by Varela et al. (1991) displays three central concerns that likewise occupy center stage in the present treatment.[38] First, Varela et al. are concerned to do justice to the active nature of perception and the way our cognitive organization reflects our physical involvement in the world. Second, they offer some powerful example of emergent behavior in simple systems.[39] Third, there is sustained attention to the notion of reciprocal (or "circular") causation and its negative implications for certain kinds of component-based reductive projects. These themes come together in the development of the idea of cognition as *enaction*. Enactive cognitive science, as Varela et al. define it, is a study of mind which does not depict cognition as the internal mirroring of an objective external world. Instead, it isolates the repeated sensorimotor interactions between agent and world as the basic locus of scientific and explanatory interest.[40]

Varela et al. are thus pursuing a closely related project to our own. There are, however, some important differences of emphasis and interest. First, Varela et al. use their reflections as evidence against realist and objectivist views of the world. I deliberately avoid this extension, which runs the risk of obscuring the scientific value of an embodied, embedded approach by linking it to the problematic idea than objects are not independent of mind.[41] My claim, in contrast, is simply that the aspects of real-world structure which biological brains represent will often be tightly geared to specific needs and sensorimotor capacities. The target of much of the present critique is thus *not* the idea that brains represent aspects of a real independent world, but rather the idea of those representations as *action-neutral* and hence as requiring extensive additional computational effort to drive intelligent responses. Second, Varela et al. (ibid., p. 9) oppose the idea that "cognition is fundamentally representation." Our approach is much more sympathetic to representationalist and information-processing analyses. It aims to partially reconceptualize ideas about the contents and formats of various inner states and processes, but not to reject the very ideas of internal representation and information-processing themselves. Finally, our treatment emphasizes a somewhat different body of cognitive scientific research (viz., the investigations of real-world robotics and autonomous-agent theory) and tries to show how the ideas and analyses emerging from this very recent research fit into the larger nexus of psychological, psychophysical, and developmental research which is the common ground of both discussions.

8.9 Minimal Representationalism

The recent skepticism concerning the role of computations and representation in cognitive science is, I believe, overblown. Much of the debate can be better cast as a discussion between the fans of maximal, detailed, action-neutral inner world models and those (including this author) who suspect that much intelligent behavior depends on more minimal resources such as multiple, partial, personalized and/or action-oriented types of inner encoding. Similarly, much of the opposition to the idea of the brain as a computing device is better cast as opposition to the idea of the brain as encoding "fully programmed specifications" of development or action. The minimal conditions under which internal-representation talk will be useful, I have argued, obtain whenever we can successfully unpack the complex causal web of influences so as to reveal the information-processing adaptive role of some system of states or of processes—a system that may involve as much spatial distribution or temporal complexity as is compatible with successfully identifying the physical configurations that stand in for specific states of affairs. Such liberalism may disturb those whose intuitions about such matters were forged in the more restrictive furnaces of reflection on language, texts, and artificial grammars,[42] but I suspect that all parties will agree that one important lesson of ongoing work in both neuroscience and dynamical systems theory is that we should not be narrow-minded about the nature of the inner events that help explain behavioral success. Such inner events may include all kinds of complex neural processes depending upon wide ranges of dynamical property, including chaotic attractors, limit cycles, potential wells, trajectories in state space, values of collective or systemic variables, and much else.[43] Stories invoking internal representation, likewise, may come to coopt such highly complex, nonlocal, dynamical processes as the vehicles of specific kinds of information and knowledge. To the extent that this occurs, the notion of internal representation itself may be subtly transformed, losing especially those classical connotations that invite us to think of relatively simple, spatially and/or temporally localized structures as the typical vehicles of representation.

There exist, to be sure, cases that pose an especially hard problem. These are cases involving processes of continuous reciprocal causation between internal and external factors. Such continuous interplay, however, appears unlikely to characterize the range of cases for which the representational approach is in any case most compelling—viz., cases involving reasoning about the distant, the nonexistent, or the highly abstract. In such cases, the focus shifts to the internal dynamics of the system under study. The crucial and still-unresolved question is whether these internal dynamics will themselves reward a somewhat more liberalized but still recognizably representation-based understanding. To pump the negative intuition, it may be suggested that, as the internal dynamics grow more and more complex, or as the putative contents grow more and more minimal (personalized, action-oriented), the explanatory leverage provided by the representational glosses must diminish, eventually vanishing beneath some as-yet-to-be-ascertained threshold. To pump the positive intuition, it may be noted that no alternative understanding of genuinely representation-hungry problem solving yet exists, and that it is hard to see how to give crisp, general, and perspicuous explanations of much of our adaptive success without somehow reinventing the ideas of complex information processing and of content-bearing inner states.

Further progress with these issues, it seems likely, must await the generation and analysis of a wider range of practical demonstrations: dynamical systems models that target reasoning and action in ever more complex and abstract domains. As such research unfolds, I think, we will see a rather delicate and cooperative coevolution between multiple types of analysis and insight. We will see the emergence of new ideas about representation and about computation—ideas that incorporate the economies of action-oriented inner states and continuous analog processing, and recognize the complex cooperative dance of a variety of internal and external sources of variance. We will learn to mark the information-processing adaptive role of inner states and processes in ways which do not blind us to the complexities of the interactive exchanges that undergird so much of our adaptive success. But, in general, we will find ourselves adding new tools to cognitive science's tool kit, refining and reconfiguring but not abandoning those we already posses. After all, if *the brain* were so simple that a single approach could unlock its secrets, *we* would be so simple that we couldn't do the job![44]

III

Further

So the Hieronymus Bosch bus headed out of Kesey's place with the destination sign in front reading "Furthur" and a sign in the back saying "Caution: Weird Load."

—Tom Wolfe, *The Electric Kool-Aid Acid Test*

We live in a world where speech is an institution.

—Maurice Merleau-Ponty, *Phenomenology of Perception* (1945/1962), p. 184

9

Minds and Markets

9.1 Wild Brains, Scaffolded Minds

Biological reason, we have seen, often consists in a rag-bag of "quick and dirty" on-line stratagems—stratagems available, in part, thanks to our ability to participate in various kinds of collective or environment-exploiting problem solving. It is natural to wonder, however, just how much leverage (if any) this approach[1] offers for understanding the most advanced and distinctive aspects of human cognition—not walking, reaching, wall following, and visual search, but voting, consumer choice, planning a two-week vacation, running a country, and so on. Do these more exotic domains at last reveal the delicate flower of logical, classical, symbol-manipulating, internal cogitation? Is it here that we at last locate the great divide between detached human reason and the cognitive profiles of other animals?[2]

In the remaining chapters, I shall tentatively suggest that there is no need to posit such a great divide, that the basic form of individual reason (fast pattern completion in multiple neural systems) is common throughout nature, and that where we human beings really score is in our amazing capacities to create and maintain a variety of special external structures (symbolic and social-institutional). These external structures function so as to complement our individual cognitive profiles and to diffuse human reason across wider and wider social and physical networks whose collective computations exhibit their own special dynamics and properties.

This extension of our basic framework to more advanced cases involves three main moves. First, individual reasoning is again cast as some kind of fast, pattern-completing style of computation. Second, substantial problem-solving work is offloaded onto external structures and processes—but these structures and processes now tend to be social and institutional rather than brute physical. And third, the role of public language (both as a means of social coordination and as a tool for individual thought) now becomes paramount.

The idea, in short, is that advanced cognition depends crucially on our abilities to *dissipate* reasoning: to diffuse achieved knowledge and practical wisdom through complex social structures, and to reduce the loads on individual brains by locating those brains in complex webs of linguistic, social, political, and institutional constraints. We thus begin to glimpse ways of confronting the phenomena of advanced cognition that are at least broadly continuous with the basic approach pursued in the simpler cases. Human brains, if this is anywhere near the mark, are not so different from the fragmented, special-purpose, action-oriented organs of other animals and autonomous robots. But we excel in one crucial respect: we are masters at structuring our physical and social worlds so as to press complex coherent behaviors from these unruly resources. We use intelligence to structure our environment so that we can succeed with *less* intelligence. Our brains make the world smart so that we can be dumb in peace! Or, to look at it another way, it is the human brain *plus* these chunks of external scaffolding that finally constitutes the smart, rational inference engine we call mind. Looked at that way, we are smart after all—but our boundaries extend further out into the world than we might have initially supposed.[3]

9.2 Lost in the Supermarket

You go into the supermarket to buy a can of beans. Faced with a daunting array of brands and prices, you must settle on a purchase. In such circumstances, the rational agent, according to classical economic theory, proceeds roughly as follows: The agent has some preexisting and comprehensive set of preferences, reflecting quality, cost, and perhaps other factors (country of origin or whatever). Such preferences have associated weights or values, resulting in a rank ordering of desired features. This complex (and consistent) preference ordering is then applied to a perfect state of knowledge about the options which the world (supermarket) offers. The bean-selecting agent then acts so as to maximize expected utility; i.e., the agent buys the item that most closely satisfies the requirements laid out in the ordered set of preferences (Friedman 1953). This image of rational economic choice has recently been termed the paradigm of *substantive rationality* (Denzau and North 1995).

Taken as a theory of the psychological mechanisms of daily individual choice, the substantive rationality model is, however, deeply flawed. The main trouble, as Herbert Simon (1982) famously pointed out, is that human brains are, at best, loci of only partial or bounded[4] rationality. Our brains, as the preceding chapters repeatedly attest, were not designed as instruments of unhurried, fully informed reason. They were not designed to yield perfect responses on the assumption of perfect information.

In view of the "quick and dirty," bounded, time-constrained nature of biological cognition, it is perhaps surprising that classical economic theory, with its vision of the fully informed, logically consistent, cool, unhurried reasoner, has done as well as it has. Why, given the gross psychological irrealism of its model of human choosing, has traditional economics yielded at least moderately successful and predictive models of, for example, the behaviors of firms (in competitive posted-price markets) and of political parties and the outcomes of experimental manipulations such as the "double auction" (Satz and Ferejohn 1994; Denzau and North 1995). And why—on a less optimistic note—has it *failed* to illuminate a whole panoply other economic and social phenomena? Among the notable failures are the failure to model large-scale economic change over time and the failure to model choice under conditions of strong uncertainty—for example, cases where there is no preexisting set of outcomes that can be rank ordered according to desirability (Denzau and North 1995; North 1993). These are fundamental failures insofar as they ramify across a wide variety of more specific cases, such as the inability to model voter behavior, the inability to predict the development of social and economic institutions, and the inability to address the bulk of the choices faced by those who make public policy.[5]

The pattern of successes and failures is both fascinating and informative, for the best explanation of the pattern appears to involve a dissociation between cases of what may be termed *highly scaffolded choice* and cases of more weakly constrained individual cogitation. The paradigm of substantive rationality, as several authors have recently argued,[6] seems to work best in the highly scaffolded case and to falter and fail as the role of weakly constrained individual cogitation increases.

The idea of highly scaffolded choice is at the heart of important recent treatments by Satz and Ferejohn (1994) and Denzau and North (1995). The common theme is that neoclassical economic theory works best in situations in which individual rational choice has been severely limited by the quasi-evolutionary selection of constraining policies and institutional practices. The irony is explicitly noted by Satz and Ferejohn: "the [traditional] theory of rational choice is most powerful in contexts where choice is limited" (p. 72). How can this be? According to Satz and Ferejohn, the reason is simple: What is doing the work in such cases is not so much the individual's cogitations as the larger social and institutional structures in which the individual is embedded. These structures have themselves evolved and prospered (in the cases where economic theory works) by promoting the selection of collective actions that do indeed maximize returns relative to a fixed set of goals. For example, the competitive environment of capital markets ensures that, by and large, only firms that maximize profits survive. It is this fact, rather than facts about the beliefs, desires, or other psychological features of the individuals involved, that ensures the frequent success of substantive rationality models in predicting the behavior of firms. Strong constraints imposed by the larger market structure result in firm-level strategies and policies that maximize profits. In the embrace of such powerful scaffolding, the particular theories and worldviews of individuals may at times make little impact on overall firm-level behavior. Where the external scaffolding of policies, infrastructure, and customs is strong and (importantly) is a result of competitive selection, the individual members are, in effect, interchangeable cogs in a larger machine. The larger machine extends way outside the individual, incorporating large-scale social, physical, and even geopolitical structures. And it is the diffused reasoning and behavior of this larger machine that traditional economic theory often succeeds in modeling. A wide variety of individual psychological profiles are fully compatible with specific functional roles within such a larger machine. As Satz and Ferejohn (ibid., p. 79) remark: "Many sets of individual motivations are compatible with the constraints that competitive market environments place on a firms behavior. In explaining firm behavior, we often confront causal patterns that hold constant across the diverse realizations of maximizing activity found in Calvinist England and the Savings and Loan Community in Texas."

In contrast, the theory of consumer behavior is weak and less successful. This is because individual worldviews and ideas loom large in consumer choice and the external scaffolding is commensurately weaker. Similarly, the theory of voting behavior is weak in comparison with the theory of party behavior in electoral competitions. Once again, the parties survive only subject to strong selection pressures that enforce vote-maximizing activity. In comparison, individual choice is relatively unconstrained (ibid., pp. 79–80).

Satz and Ferejohn suggest that the crucial factor distinguishing the successful and unsuccessful cases (of the use of neoclassical, substantive-rationality-assuming theory) is the availability of a structurally determined *theory of interests*. In cases where the overall structuring environment acts so as to select in favor of actions which are restricted so as to conform to a specific model of preferences, neoclassical theory works. And it works because individual psychology no longer matters: the "preferences" are imposed by the wider situation and need not be echoed in individual psychology. For example, in a democratic, two-party electoral system the overall situation selects for the party that acts to maximize votes. This external structuring force allows us to impute "preferences" on the basis of the constraints on success in such a larger machine. The constraints on individual voters are much weaker. Hence, real psychological profiles come to the fore, and neoclassical theory breaks down (Satz and Ferejohn 1994, pp. 79-80; North 1993, p. 7). This general diagnosis is supported by the analysis of Denzau and North (1995). They note that traditional economic theory nicely models choice in competitive posted price markets and in certain restricted experimental studies. In such cases, they suggest, certain institutional features play major roles in promoting "maximizing-style" economic performance. By way of illustration, Denzau and North cite some fascinating computational studies by Gode and Sunder (1992) that invoke "zero-intelligence" traders—simulated agents who do not actively theorize, recall events, or try to maximize returns. When such simple agents were constrained to bid only in ways that would not yield immediate losses, an efficiency of 75 percent (measured as "the percentage of sum of potential buyer and seller rents" (ibid., p. 5) was achieved. Replacing the zero-intelligence (ZI) traders with humans increased efficiency by a mere 1 percent. But altering the institutional scaffolding (e.g. from collecting all bids in a double auction before contracting to allowing simultaneous bidding and contracting) yielded a 6 percent improvement in efficiency. The strong conclusion is that "most efficiency gains in some resource allocation situations may be attributed to institutional details, independent of their effects on rational traders" (ibid., p. 5).

The results of the ZI-trader experiments clearly demonstrate the power of institutional settings and external constraints to promote collective behaviors that conform to the model of substantive rationality. Such results fit nicely with the otherwise disquieting news that the bulk of traditional economics would be unaffected if we assumed that individuals chose randomly (Alchian 1950, cited in Satz and Ferejohn 1994) rather than by maximizing preferences, and that pigeons and rats can often perform in ways consistent with the theory of substantive rationality (Kagel 1987, cited in Satz and Ferejohn 1994). Such results make sense if the scaffolding of choice by larger-scale constraining structures is sometimes the strongest carrier of maximizing force. In the extreme limiting cases of such constraint, the individual chooser is indeed a mere cog—a constrained functional role played as well by a zero-intelligence trader, a pigeon, a rat, a human trader, or, in the worst cases, a coin-flipping device.[7]

9.3 The Intelligent Office?

The moral so far is that the scaffolding matters: the external structuring provided by institutions and organizations bears much of the explanatory burden for explaining current economic patterns. To see where human psychology fits in, let us begin by asking: What kind of individual mind *needs* an external scaffold?

A vital role for external structure and scaffolding is, as we have seen, strongly predicted by recent work on individual cognition. Simon's (1982) notion of bounded rationality was probably the first step in this direction.

But although Simon rightly rejected the view of human agents as perfect logical reasoners, he remained committed to a basically classicist model of computation (see introduction and chapter 3 above) as involving explicit rules and quasi-linguistic data structures. The major difference was just the use of heuristics, with the goal of *satisficing* rather than optimizing—i.e., the use of "rules of thumb" to find a workable solution with minimal expenditures of time and processing power.

The reemergence of connectionist (artificial neural networks, parallel distributed processing) ideas (see chapter 3 above) took us farther by challenging classical models of internal representation and of computational process.

We saw in section 3.3 how such systems in effect substitute fast pattern recognition for step-by-step inference and reasoning. This substitution yields a particular profile of strengths (motor skills, face recognition, etc.) and weaknesses (long-term planning, logic)—one that gives us a useful fix on the specific ways in which external structures may complement and augment bare individual cognition. The external structures, it was argued, enable us to negotiate problem domains that require the sequential and systematic deployment of basic pattern-completing capacities and the presentation and reuse of intermediate results. The simple example rehearsed in chapter 3 concerned the use of pen and paper to amplify simple arithmetical knowledge (e.g. that $7 \times 7 = 49$) into solutions to more complex problems (e.g. 777×777). We can now see, in barest outline, how institutions, firms, and organizations seem to share many of the key properties of pen, paper, and arithmetical practice in this example. Pen and paper provide an external medium in which we behave (using basic on-line resources) in ways dictated by the general policy or practice of long multiplication. Most of us do not know the mathematical justification of the procedure. But we use it, and it works. Similarly, firms and organizations provide an external resource in which individuals behave in ways dictated by norms, policies, and practices. Daily problem solving, in these arenas, often involves locally effective pattern-recognition strategies which are invoked as a result of some externally originating prompt (such as a green slip in the "in" tray, discharged in a preset manner) and which leave their marks as further traces (slips of paper, e-mail messages, whatever) which then are available for future manipulations within the overarching machinery of the firm. In these contexts, in the short term at least, the role of individual rationality can become somewhat marginal. If the overall machinery and strategies have been selected so as to maximize profits, the fact that the individuals are cogs deploying very bounded forms of pattern-completing rationality will not matter. (Individual neurons are, if you like, even more restricted cogs, but once organized into brains by natural selection they too support a grander kind of reason.)

Much of what goes on in the complex world of humans may thus, somewhat surprisingly, be understood as involving something rather akin to the "stigmergic algorithms" introduced in section 4.3. Stigmergy, recall, involves the use of external structures to control, prompt, and coordinate individual actions. Such external structures can themselves be acted upon and thus mold future behaviors in turn. In the case of termite nest building, the actions of individual termites were controlled by local nest structure yet often involved modifications of that structure which in turn prompted further activity by the same or other individuals. Humans, even when immersed in the constraining environments of large social political or economical institutions, are, of course, not termites! Unlike the termite, we will not always perform an action simply because an external prompt seems to demand it. However, our collective successes (and sometimes our collective failures) may often be best understood by seeing the individual as choosing his or her responses only within the often powerful constraints imposed by the broader social and institutional contexts of action. And this, indeed, is just what we should expect once we recognize that the computational nature of individual cognition is not ideally suited to the negotiation of certain types of complex domains. In these cases, it would seem, we solve the problem (e.g. building a jumbo jet or running a country) only indirectly—by creating larger external structures, both physical and social, which can then prompt and coordinate a long sequence of individually tractable episodes of problem solving, preserving and transmitting partial solutions along the way.

9.4 Inside the Machine

Organizations, factories, offices, institutions, and such are the large-scale scaffolds of our distinctive cognitive success. But as surely as these larger wholes inform and scaffold individual thought, they themselves are structured and informed by the communicative acts of individuals and by episodes of solitary problem solving. One crucial project for the cognitive sciences of the embodied mind is to begin the hard task of understanding and analyzing this complex reciprocal relationship—a daunting task that will require the use of simulations which operate at multiple time scales and levels of organization. Such simulations should ideally encompass genetic evolutionary change, individual learning and problem solving, processes of cultural and artifactual evolution, and the emergent problem-solving capacities of groups of communicating agents. This is, alas, a little too much to ask, given the current state of the art. But it is at least possible to begin to scratch the surface of the issues.

There have been some nice attempts at modeling something of the interplay between genetic evolution and individual learning (Ackley and Littman 1992; Nolfi and Parisi 1991; see discussions in Clark 1993 and Clark (to appear)). Of more relevance to the present discussion, however, are efforts to model the interplay of individual learning, cultural and artifactual evolution, and patterns of inter-group communication. In this vein, Hutchins (1995) set out to investigate how various patterns of communication affect the collective problem-solving capacities of small groups of simple artifactual "agents." Each agent, in this simulation, was a small neural network comprising a few linked processing units. Each unit coded for some specific environmental feature. Excitatory links connected mutually supportive features; inhibitory links connected mutually inconsistent features. For example, a feature like "is a dog" would be coded by a single unit with excitatory links to (e.g.) "barks" and "has fur" units and inhibitory links to (e.g.) "meows" and "is a cat" units (the latter being themselves linked by an excitatory connection). Such networks are known as *constraint-satisfaction networks*.

Once a constraint-satisfaction network is set up (either by learning or by hand coding), it exhibits nice properties of pattern-completion-style reasoning. Thus, imagine that the various units receive input signals from the environment. Activation of a few units that figure in a linked web of excitatory connections will yield activity across all the other linked units. The input "barks" will thus yield a global activation profile appropriate to the category "dog," and so on. Individual units often "choose" whether or not to respond (become active) by summing the inputs received along various channels and comparing the result to some threshold level. As a result, once a constraint-satisfaction network settles into an interpretation of the input (e.g. by having all the dog-feature units become active), dislodging it becomes difficult because the units lend each other considerable mutual support. This feature of such networks, Hutchins points out, corresponds rather nicely to the familiar psychological effect of confirmation bias—viz., the tendency to ignore, discount, or creatively reinterpret evidence (such as a solitary "meows" input) that goes against some hypothesis or model that is already in place. (See, e.g., Wason 1968.)

Now imagine a community of constraint-satisfaction networks in which each network has a different initial activity level ("predisposition") and different access to environmental data. Hutchins shows that in such cases the precise way in which the inter-network communication is structured makes a profound difference to the kind of collective problem solving displayed. Surprisingly, Hutchins (p. 252) found that in such cases more communication is not always better than less. In particular, if from the outset all the networks are allowed to influence the activity of the others (to communicate), the overall system shows an extreme degree of confirmation bias—much more than any one of the individual nets studied in isolation. The reason is that the dense communication patterns impose a powerful drive to rapidly discover a shared interpretation of the data—to find a stable pattern of activity across all the units. The individual nets, instead of giving due weight to the external input data, focus more heavily on these internal constraints (the need to find a set of activation patterns that doesn't disrupt the others). As a result, the social group rushes "to the interpretation that is closest to the center of gravity of their predispositions, regardless of the evidence" (ibid., p. 259).

By contrast, if you restrict the level of early communication, this gives each individual network time to balance its own predispositions against the environmental evidence. If inter-network communication is *subsequently* enabled, then overall confirmation bias is actively reduced—that is, the group is more likely than the average member to fix on a correct solution. Such results suggest that the collective advantage of a jury over an individual decision may dissipate proportionally to the level of early communication between members.[8] More important, however, the example illustrates one way in which we may begin to understand, in a rigorous manner, some aspects of the delicate interplay between individual cognition and group-level dynamics. Such understanding will surely be crucial to a better appreciation of the roles of institutional and organizational structures in determining collective problem solving, and of the balance between individual cognition and the external scaffolding which it both shapes and inhabits.

One moral of the simple demonstration just rehearsed is that there is scope for patterns of inter-agent communication to evolve (over cultural-evolutionary time) so as to better serve the problem-solving needs of a given collective. In a fascinating earlier simulation, Hutchins and Hazelhurst (1991) showed that the cultural artifacts (words and symbols) that flow around inside the collective machine are themselves capable of "evolving" so as to better serve specific problem-solving needs. In this study Hutchins (a cognitive scientist) and Hazelhurst (a cultural anthropologist) created a simple computer simulation in which successive generations of simple connectionist networks gradually improved their problem-solving capacity by creating and passing on a set of cultural artifacts —viz., a simple language encoding information about some salient correlations among environmental events. The simulation involved a group of "citizens" (connectionist nets) able to learn from two kinds of environmental structure: "natural structure" (observed correlations between events—in this case, between phases of the moon and states of the tide) and "artifactual structure" (learning by exposure to symbols representing the states of moon and tide). The nets are able, by virtue of the usual pattern-completing and learning abilities, to learn to associate symbols with events, and to denote events with symbols. They are thus able to generate symbols to reflect experiences, and to use symbols to *cause* the kinds of experience which the real-world event (itself just another kind of coding in this simple simulation) would usually bring about. Exposure to symbols thus causes a kind of "vicarious experience" of the associated events. In addition, some simulations incorporated an "artifact selection bias" in which cultural products (the symbol structures) were selected by other citizens with a probability based in part on the competence (degree of success) of the net that produced them.

The Hutchins-Hazelhurst study involved observing the relative success of many generations of networks. But, in contrast with the genetic-algorithm work discussed in chapter 5, the subsequent generations were identical in internal structure—no genetic improvements were allowed. Nonetheless, the gradual accumulation of better external artifacts (the symbolic structures representing moon and tide states) enabled later generations to learn environmental regularities that earlier ones could not learn. The contribution of each individual to future generations was not genetic; rather, it consisted of a symbolic artifact comprising entries for phases of the moon and states of the tide. Citizens of subsequent generations were trained in part on the artifacts of their immediate predecessors, with selection of such artifacts either made at random (all artifacts of the previous generation equally likely to be used) or relative to the selection bias (thus favoring the better artifacts).

The results were clear: Early generations could not predict the regularity. Later generations, identical at birth and using the same learning procedures, were able to solve the problem. Simulations involving the selection bias were more successful than those based on random choice. The existence of artifactual products and strategies of artifactual selection thus enables a kind of multi-generational learning which is independent of genetic change and which greatly expands the horizons of individual learning.[9]

We are seeing in these simple simulations some of the first attempts to put quantitative, analytic flesh on ideas about collective problem solving in communities of agents capable of creating and exploiting various kinds of external symbol structure. These symbol structures are the lifeblood flowing through the larger social and institutional machinery that both molds and empowers individual human thought.

9.5 Designer Environments

Rodney Brooks, the creator of many of the mobile robots described in the preceding chapters, recently asked this question: How can we get coherent behavior from multiple adaptive processes without centralized control? The question is pressing if, as many roboticists and neuroscientists suspect, even advanced human cognition depends on multiple inner systems, with limited communication, exploiting partial and action-oriented forms of internal representation. Without the great central homunculus—the inner area in which, as Dennett (1991) puts it, everything "comes together"—what stops behavior from becoming chaotic and self-defeating? Brooks (1994) considers three sources of constraint: natural coherence (where the physical world determines, e.g., that action A will be performed before action B), designed coherence (where the system has, e.g., a built-in hierarchy of goals), and various forms of cheap global modulation (such as hormonal effects).

To this list we can now add the idea of stigmergic self-modulation: the process by which intelligent brains *actively* structure their own external (physical and social) worlds so as to make for successful actions with less individual computation. The coherence and the problem-solving power of much human activity, it seems, may be rooted in the simple yet often-ignored fact that we are the most prodigious creatures and exploiters of external scaffolding on the planet. We build "designer environments" in which human reason is able to far outstrip the computational ambit of the unaugmented biological brain. Advanced reason is thus above all the realm of the *scaffolded* brain: the brain in its bodily context, interacting with a complex world of physical and social structures. These external structures both constrain and augment the problem-solving activities of the basic brain, whose role is largely to support a succession of iterated, local, pattern-completing responses. The successes of classical economics (to take just one example) emerge, within this paradigm, as depending largely on the short-term dynamics of responses strongly determined by particular kinds of institutional or organizational structures: structures which have *themselves* evolved as a result of selective pressure to maximize rewards of a certain kind.

Nonetheless, these external scaffoldings are, in most cases, themselves the products of individual and collective human thought and activity. The present discussion thus barely scratches the surface of a large and difficult project: understanding the way our brains both structure and inhabit a world populated by cultures, countries, languages, organizations, institutions, political parties, e-mail networks, and all the vast paraphernalia of external structures and scaffoldings which guide and inform our daily actions.

All of this, as Hutchins (1995) pointedly notes, serves only to remind us of what we already knew: if our achievements exceed those of our forebears, it isn't because our brains are any smarter than theirs. Rather, our brains are the cogs in larger social and cultural machines—machines that bear the mark of vast bodies of previous search and effort, both individual and collective. This machinery is, quite literally, the persisting embodiment of the wealth of achieved knowledge. It is this leviathan of diffused reason that presses maximal benefits from our own simple efforts and is thus the primary vehicle of our distinctive cognitive success.

10

Language: The Ultimate Artifact

10.1 Word Power

What does public language do for us? There is a common, easy answer, which, though not incorrect, is subtly misleading. The easy answer is that language helps us to communicate ideas. It lets other human beings profit from what we know, and it enables us to profit from what they know. This is surely true, and it locates one major wellspring of our rather unique kind of cognitive success. However, the emphasis on language as a medium of communication tends to blind us to a subtler but equally potent role: the role of language as a tool[1] that alters the nature of the computational tasks involved in various kinds of problem solving.

The basic idea is simple enough. Consider a familiar tool or artifact, say a pair of scissors.[2] Such an artifact typically exhibits a kind of double adaptation—a two-way fit, both to the user and to the task. On the one hand, the shape of the scissors is remarkably well fitted to the form and the manipulative capacities of the human hand. On the other hand (so to speak), the artifact, when it is in use, confers on the agent some characteristic powers or capacities which humans do not naturally possess: the ability to make neat straight cuts in certain papers and fabrics, the ability to open bubble packs, and so forth. This is obvious enough; why else would we value the artifact at all?

Public language is in many ways the ultimate artifact. Not only does it confer on us added powers of communication; it also enables us to reshape a variety of difficult but important tasks into formats better suited to the basic computational capacities of the human brain. Just as scissors enable us to exploit our basic manipulative capacities to fulfill new ends, language enables us to exploit our basic cognitive capacities of pattern recognition and transformation in ways that reach out to new behavioral and intellectual horizons. Moreover, public language may even exhibit the kind of double adaptation described above, and may hence constitute a body of linguistic artifacts whose form is itself in part evolved so as to exploit the contingencies and biases of human learning and recall. (This reverse adaptation—of the artifact to the user—suggests a possible angle on the controversy concerning innate mechanisms for language acquisition and understanding.) Finally, the sheer intimacy of the relations between human thought and the tools of public language bequeaths an interesting puzzle. For in this case, especially, it is a delicate matter to determine where the user ends and the tool begins!

10.2 Beyond Communication

The idea that language may do far more than merely serve as a vehicle for communication is not new. It is clearly present in the work of developmentalists such as Lev Vygotsky (1962) and Laura Berk (see, e.g., Diaz and Berk 1992). It figures in the philosophical conjectures and arguments of, e.g., Peter Carruthers (to appear) and Ray Jackendoff (to appear). And it surfaces in the more cognitive-science-oriented speculations of Daniel Dennett (1991, 1995). It will be helpful to review some of the central ideas in this literature before pursuing our preferred version—viz., the idea of language as a computational transformer that allows pattern-completing brains to tackle otherwise intractable classes of cognitive problems.

In the 1930s, Vygotsky, a psychologist, pioneered the idea that the use of public language had profound effects on cognitive development. He posited powerful links among speech, social experience, and learning. Two Vygotskian ideas that are especially pertinent for present purposes concern private speech and scaffolded action (action within the "zone of proximal development"—see Vygotsky 1962 and chapter 3 above). We have called an action "scaffolded" to the extent that it relies on some kind of external support. Such support could come from the use of tools or from exploitation of the knowledge and skills of others; that is to say, scaffolding (as I shall use the term[3]) denotes a broad class of physical, cognitive, and social augmentations—augmentations that allow us to achieve some goal that would otherwise be beyond us. Simple examples include the use of a compass and a pencil to draw a perfect circle, the role of other crew members in enabling a ship's pilot to steer a course, and an infant's ability to take its first steps only while suspended in the enabling grip of its parents. Vygotsky's focus on the "zone of proximal development" was concerned with cases in which a child is temporarily able to succeed at designated tasks only by courtesy of the guidance or help provided by another human being (usually a parent or a teacher), but the idea dovetails with Vygotsky's interest in private speech in the following way: When a child is "talked through" a tricky challenge by a more experienced agent, the child can often succeed at a task that would otherwise prove impossible. (Think of learning to tie your shoelaces.) Later, when the adult is absent, the child can conduct a similar dialogue, but this time with herself. But even in this latter case, it is argued, the speech (be it vocal or "internalized") functions so as to guide behavior, to focus attention, and to guard against common errors. In such cases, the role of language is to guide and shape our own behavior—it is a tool for structuring and controlling action, not merely a medium of information transfer between agents.

This Vygotskian image is supported by more recent bodies of developmental research. Berk and Garvin (1984) observed and recorded the ongoing speech of a group of children between the ages of 5 and 10 years. They found that most of the children's private speech (speech not addressed to some other listener) seemed keyed to the direction and control of the child's own actions, and that the incidence of such speech increased when the child was alone and trying to perform some difficult task. In subsequent studies (Bivens and Berk 1990; Berk 1994) it was found that the children who made the greatest numbers of self-directed comments were the ones who subsequently mastered the tasks best. Berk concluded, from these and other studies, that self-directed speech (be it vocal or silent inner rehearsal) is a crucial cognitive tool that allows us to highlight the most puzzling features of new situations and to better direct and control our own problem-solving actions.

The theme of language as a tool has also been developed by the philosopher Christopher Gauker. Gauker's concern, however, is to rethink the intra-individual role of language in terms of what he calls a "cause-effect analysis." The idea here is to depict public language "not as a tool for representing the world or expressing ones thoughts but a tool for effecting changes in one's environment" (Gauker 1990, p. 31). To get the flavor of this, consider the use of a symbol by a chimpanzee to request a banana. The chimp touches a specific key on a keypad (the precise physical location of the key can be varied between trials) and learns that making *that* symbol light tends to promote the arrival of bananas. The chimp's quasi-linguistic understanding is explicable, Gauker suggests, in terms of the chimp's appreciation of a cause-effect relationship between the symbol production and changes in its local environment. Gauker looks at a variety of symbol-using behaviors and concludes that they all succumb to this kind of analysis. This leads him to hypothesize that, although clearly more complex, human beings' linguistic understanding likewise "consists in a grasp of the causal relations into which linguistic signs may enter" (ibid., p. 44).

Gauker tends to see the role of language as, if you like, directly causal: as a way of getting things done, much like reaching out your hand and grabbing a cake. However, the idea that we learn, by experience, of the peculiar causal potencies of specific signs and symbols is, in principle, much broader. We might even, as in the Vygotskian examples, discover that the self-directed utterance of words and phrases has certain effects on our own behavior.[4] We might also learn to exploit language as a tool in a variety of even less direct ways, as a means of altering the shape of computational problem spaces (see section 10.3).

One obvious question raised by the putative role of language as a self-directed tool is "How does it work?" What is it about, for example, self-directed speech that fits it to play a guiding role? After all, it is not at all clear how we can tell ourselves anything we don't already know! Surely all public language can ever be is a medium for expressing ideas already formulated and understood in some other, more basic inner code. This is precisely the view that a supra-communicative account of language ultimately has to reject. One way to reject it is to depict public language as itself the medium of a special kind of thought. Another (by no means exclusive, and not altogether distinct) way is to depict linguaform inputs as having distinctive *effects* on some inner computational device. Carruthers (to appear) champions the first of these; Dennett (1991) offers a version of the second.[5] Carruthers argues that, in this case at least, we should take very seriously the evidence of our own introspection. It certainly often seems as if our very thoughts are composed of the words and sentences of public language. And the reason we have this impression, Carruthers argues, is because it is true: ". . . inner thinking is literally done in inner speech."[6] By extension, Carruthers is able to view many uses of language as less a matter of simple communication than a matter of what he nicely terms *public thinking*. This perspective fits satisfyingly with the Vygotskian view championed by Berk and is also applicable to the interesting case of writing down our ideas. Carruthers (ibid., p. 56) suggests that "one does not *first* entertain a private thought and *then* write it down: rather, the thinking *is* the writing." I shall return to this point later (see section 10.3 and the epilogue), since I believe that what Carruthers says is almost right but that we can better understand the kind of case he has in mind by treating the writing as an environmental manipulation that transforms the problem space for human brains.

A further way to unpack a supra-communicative view of language, as has been noted, is to suppose that the linguistic inputs actually reprogram or otherwise alter the high-level computational structure of the brain itself. The exegesis is delicate (and therefore tentative), but Dennett (1991, p. 278) seems to hold such a view when he suggests that "conscious human minds are more-or-less serial virtual machines implemented-inefficiently-on the parallel hardware that evolution has provided for us." In this and other passages of the same work, the idea seems to be that the bombardment of (something like) parallel-processing, connectionist, pattern-completing brains by (among other things) public-language texts and sentences (reminders, plans, exhortations, questions, etc.) results in a kind of cognitive reorganization akin to that which occurs when one computer system simulates another. In such cases, the installation of a new program allows the user to treat a serial LISP machine (for example) as if it were a massively parallel connectionist device. What Dennett is proposing is, he tells us (ibid., p. 218), the same trick in reverse—the simulation of something like a serial logic engine using the altogether different resources of the massively parallel neural networks that biological evolution rightly favors for real-world, real-time survival and action.

Strikingly, Dennett (1995, pp. 370–373) suggests that it is this subtle reprogramming of the brain by (primarily) linguistic bombardment that yields the phenomena of human consciousness (our sense of self) and enables us to far surpass the behavioral and cognitive achievements of most other animals. Dennett thus depicts our advanced cognitive skills as attributable in large part not to our innate hardware (which may differ only in small, though important, ways from that of other animals) but to the special way that various plastic (programmable) features of the brain are modified by the effects of culture and language. As Dennett (1991, p. 219) puts it, the serial machine is installed by courtesy of "myriad microsettings in the plasticity of the brain." Of course, mere exposure to culture and language is not sufficient to ensure human-like cognition. You can expose a cockroach to all the language you like and get no trace of the cognitive transformations Dennett sees in us. Dennett's claim is not that there are *no* initial hardware-level differences. Rather it is that some relatively small hardware differences (e.g. between humans and chimpanzees) allow us to both create and benefit from public language and other cultural developments in ways that lead to a great snowball of cognitive change and augmentation—including, perhaps, the literal installation of a new kind of computational device inside the brain.

Dennett's vision is complex and not altogether unambiguous. The view I want to develop is clearly deeply related to it, but it differs (I think) in one crucial respect. Whereas Dennett sees public language as both a cognitive tool and a source of some profound but subtle reorganization of the brain, I am inclined to see it as in essence just a tool—an external resource that complements but does not profoundly alter the brain's own basic modes of representation and computation. That is to say, I see the changes as relatively superficial ones geared to allowing us to use and exploit various *external* resource to the full. The positions are not, of course, wholly distinct. The mere fact that we often mentally rehearse sentences in our heads and use these to guide and alter our behavior means that one cannot and should not treat language and culture as wholly external resources. Nonetheless, it remains possible that such rehearsal does not involve the use of any fundamentally different kind of computational device in the brain so much as the use of the same old (essentially pattern-completing) resources to model the special kinds of behavior observed in the world of public language. And, as Paul Churchland (1995, pp. 264–269) points out, there is indeed a class of connectionist networks ("recurrent networks"—see chapter 7 above, Elman 1993, and further discussion in Clark 1993) that seem well suited to modeling and supporting such linguistic behavior.

This view of inner rehearsal is nicely developed by the connectionists David Rumelhart, Paul Smolensky, James McClelland, and Geoffrey Hinton, who argue that the general strategy of "mentally modeling" the behavior of selected aspects of our environment is especially important insofar as it allows us to imagine external resources with which we have previously physically interacted, and to replay the dynamics of such interactions in our heads. Thus experience with drawing and using Venn diagrams allows us to train a neural network which subsequently allows us to manipulate imagined Venn diagrams in our heads. Such imaginative manipulations require a specially trained neural resource, to be sure, but there is no reason to suppose that such training results in the installation of a different *kind* of computational device. It is the same old process of pattern completion in high-dimensional representational spaces, but applied to the special domain of a specific kind of *external* representation. Rumelhart et al., who note the clear link with a Vygotskian image, summarize their view as follows (1986, p. 47):

> We can be instructed to behave in a particular way. Responding to instructions in this way can be viewed simply as responding to some environmental event. We can also remember such an instruction and "tell ourselves" what to do. We have, in this way, internalized the instruction. We believe that the process of following instructions is essentially the same whether we have told ourselves or have been told what to do. Thus even here we have a kind of internalization of an external representational format.

The larger passage (pp. 44–48) from which the above is extracted is remarkably rich and touches on several of our major themes. Rumelhart et al. note that such external formalisms are especially hard to invent and slow to develop and are themselves the kinds of product that (in an innocently bootstrapping kind of way) can evolve only thanks to the linguistically mediated processes of cultural storage and gradual refinement over many lifetimes. They also note that by using real external representations we put ourselves in a position to use our basic perceptual and motor skills to separate problems into parts and to attend to a series of subproblems, storing intermediate results along the way—an important property to which we shall return in section 10.3.

The tack I am about to pursue likewise depicts language as an external artifact designed to complement rather than transfigure the basic processing profile we share with other animals. It does not depict experience with language as a source of profound inner reprogramming. Whether it depicts inner linguistic rehearsal as at times literally constitutive of specific human cognizings (as Carruthers claims) is moot. What matters, I think, is not to try to confront the elusive question "Do we actually think in words?" (to which the answer is surely "In a sense yes and in a sense no!"), but to try to see just what computational benefits the pattern-completing brain may press from the rich environment of manipulable external symbolic structures. Time, then, to beard language in its den.

10.3 Trading Spaces

How might linguistic artifacts complement the activity of the pattern-completing brain? One key role, I suggest, is captured by the image of *trading spaces*: the agent who exploits external symbol structures is trading culturally achieved *representation* against what would otherwise be (at best) time-intensive and labor-intensive internal *computation*. This is, in fact, the very same tradeoff we often make purely internally when we stop short of actually manipulating external symbols but instead use our internal models of those very symbols to cast a problem in a notational form that makes it easier to solve. And, as has often been remarked, it is surely our prior experiences with the manipulations of real external symbols that prepares the way for these more self-contained episodes of symbolically simplified problem solving.

Examples are legion, and they include the use of the Arabic numeral system (rather than, e.g., roman numerals) as a notation for arithmetic problem solving; the use of Venn diagrams for solving problems of set theory; the use of the specialized languages of biology, physics, and so on to set up and solve complex problems; and the use of lists and schedules as aides to individual planning and group coordination. All these cases share an underlying rationale, which is to build some of the knowledge you need to solve a problem directly into the resources you use to represent the problem in the first place. But the precise details of how the tradeoff is achieved and in what ways it expands our cognitive potential vary from case to case. It is useful, then, to distinguish a variety of ways in which we may trade culturally transmitted representation against individual computational effort.

The very simplest cases are those that involve the use of external symbolic media to offload memory onto the world. Here we simply use the artifactual world of texts, diaries, notebooks, and the like as a means of systematically storing large and often complex bodies of data. We may also use simple external manipulations (such as leaving a note on the mirror) to prompt the recall, from onboard biological memory, of appropriate information and intentions at the right time. Thus, this use of linguistic artifacts is perfectly continuous with a variety of simpler environmental manipulations, such as leaving an empty olive oil bottle by the door so that you cannot help but run across it (and hence recall the need for olive oil) as you set out for the shops.

A slightly more complex case (Dennett 1993) concerns the use of labels as a source of environmental simplification. One idea here is that we use signs and labels to provide perceptually simple clues to help us negotiate complex environments. Signs for cloakrooms, for nightclubs, and for city centers all fulfill this role. They allow a little individual learning to go a very long way, helping others to find their targets in new locales without knowing in advance what, in detail, to seek or even where exactly to seek it. McClamrock (1995, p. 88) nicely describes this strategy as one in which we "enforce on the environment certain kinds of stable properties that will lessen our computational burdens and the demands on us for inference."

Closely related, but perhaps less obvious, is the provision, by the use of linguistic labels, of a greatly simplified *learning* environment for important concepts—a role already exemplified and discussed in the treatment of the Hutchins's "moon and tide" simulation in chapter 9. The use of simple labels, it seems, provides a hefty clue for the learning device, allowing it to shrink enormous search spaces to manageable size.[7]

More sophisticated benefits of the use of linguistic representation cluster around the use of language in coordinating action. We say to others that we will be at a certain place at a certain time. We even play this game with ourselves, perhaps by writing down a list of what we will do on what days. One effect of such explicit planning is to facilitate the *coordination* of actions. Thus, if another person knows you have said you'll be at the station at 9:00 A.M., they can time their taxi ride accordingly. Or, in the solo case, if you have to buy paint before touching up your car, and if you have to go to the shops to buy other items anyway, you can minimize your efforts and enforce proper sequencing by following an explicit plan. As the space of demands and opportunities grows, it often becomes necessary to use pencil and paper to collect and repeatedly reorganize the options, and then to preserve the result as a kind of external control structure available to guide your subsequent actions.

Such coordinative functions, thought important, do not exhaust the benefits of explicit (usually language-based) planning. As Michael Bratman (1987) has pointed out, the creation of explicit plans may play a special role in reducing the on-line cognitive load on resource-limited agents like ourselves. The idea here is that our plans have a kind of stability that pays dividends by reducing the amount of on-line deliberation in which we engage as we go about much of our daily business. Of course, new information can, and often does, cause us to revise our plans. But we do not let every slight change prompt a reassessment of our plans—even when, other things being equal, we might now choose slightly differently. Such stability, Bratman suggests, plays the role of blocking a wasteful process of continual reassessment and choice (except, of course, in cases where there is some quite major payoff for the disruption).[8] Linguistic exchange and formulation thus plays a key role in coordinating activities (at both inter-personal and intra-personal levels) *and* in reducing the amount of daily on-line deliberation in which we engage.

Closely related to these functions of control and coordination is the fascinating but ill-understood role of inner rehearsal of speech in manipulating our own attention and guiding our allocation of cognitive resources. The developmental results mentioned in section 10.2 (concerning the way self-directed speech enhances problem solving) suggest an image of inner speech as an extra *control loop* capable of modulating the brain's use of its own basic cognitive resources. We see such a phenomenon in inter-personal exchange when we follow written instructions, or when we respond to someone else's vocal prompts in learning to drive or to windsurf. When we practice on our own, the mental rehearsal of these same sentences acts as a controlling signal that somehow helps us to monitor and correct our own behaviors.

Dreyfus and Dreyfus (1990) have argued that inner rehearsal plays this role only in novice performance and that real experts leave behind such linguistic props and supports. But although it is clearly true that, for example, expert drivers no longer mentally rehearse such prompts as "mirror–signal–maneuver," this does not show that language-based reason plays no role at all at the expert level. An interesting recent study by Kirsh and Maglio (see chapter 3 above) concerns the roles of reaction and linguaform reflection in expert performance at the computer game Tetris. Tetris, recall, is a game in which the player attempts to accumulate a high score by the compact placement of geometric objects (zoids) which fall down from the top of the screen. As a zoid descends, the player can manipulate its fall by rotating it at the resting point of its current trajectory. When a zoid comes to rest, a new one appears at the top of the screen. The speed of fall increases with score. But (the saving grace) a full row (one in which each screen location is filled by a zoid) disappears entirely. When the player falls behind in zoid placement and the screen fills up so that new zoids cannot enter it, the game ends. Advanced play thus depends crucially on fast decision making. Hence, Tetris provides a clear case of domain in which connectionist, pattern-completion style reasoning looks required for expert performance. If the model of Dreyfus and Dreyfus is correct, moreover, such parallel, pattern-completion-style reasoning should exhaustively explain expert skill. But, interestingly, this does not seem to be so. Instead, expert play seems to depend on a delicate and non-obvious interaction between a fast, pattern-completion module and a set of explicit higher-level concerns or normative policies. The results are preliminary, and it would be inappropriate to report them in detail, but the key observation is that true Tetris experts report that they rely not solely on a set of fast adaptive responses produced by (as it were) a trained-up network but also on a set of high-level concerns or policies, which they use to monitor the outputs of the skilled network so as to "discover trends or deviations from . . . normative policy" (Kirsh and Maglio 1991, p. 10). Examples of such policies include "don't cluster in the center, but try to keep the contour flat" and "avoid piece dependencies" (ibid., pp. 8–9) Now, on the face of it, these are just the kind of rough and ready maxims we might (following Dreyfus and Dreyfus) associate with novice players only. Yet attention to these normative policies seems to mark especially the play of real experts. Still, we must wonder how such policies can help at the level of expert play, given the time constraints on responses. There is just no time for reflection on such policies to override on-line output for a given falling zoid.

It is here that Kirsh and Maglio make a suggestive conjecture. The role of the high-level policies, they suggest, is probably indirect. Instead of using the policy to override the output of a trained-up network, the effect may be to alter the focus of attention for subsequent inputs. The ideas is that the trained-up network (or "reactive module," as Kirsh and Maglio put it) will sometimes make moves that lead to dangerous situations in which the higher-level policies are not reflected. The remedy is not to override the reactive module, but to thereafter manipulate the inputs it receives so as to present feature vectors that, when processed by the reactive model in the usual way, will yield outputs in line with policy. As Kirsh and Maglio describe it, the normative policies are thus the business of a distinct and highly "language-infected" resource that indirectly modulates the behavior of a more basic, fast and fluent reactive agency. Just how this indirect modulation is accomplished is, alas, left uncomfortably vague, but Kirsh and Maglio speculate that it might work by biasing perceptual attention toward certain danger regions or by increasing the resolution of specific visual routines.

The most obvious benefit of the linguistic encoding of thoughts and ideas, is, of course, that such encoding formats our ideas into compact and easily transmitted signals that enable other human beings to refine them, to critique them, and to exploit them. This is the *communicative* role, which, I have suggested, tends to dominate our intuitive ideas about the role and function of language. But our conception even of this familiar role remains impoverished until we see that role in the specific computational context provided by broadly connectionist models of the biological brain, for one notable feature of such models is the extreme *path dependence* of their learning routines. For example, a compelling series of experiments by Jeff Elman (1994) and others showed that connectionist learning is heavily dependent on the sequence of training cases. If the early training goes wrong, the network is often unable to recover. A specific network proved able to learn complex grammatical rules from a corpus of example sentences only if it had previously been trained on a more basic subset of the examples highlighting (e.g.) verb-subject number agreement. Early exposure to the other, more complex grammatical cases (such as long-distance dependences) would lead it into bad early "solutions" (local minima) from which it was then unable to escape.[9] Human learning, like learning in artificial neural networks, appears to be hostage to at least some degree of path dependence. Certain ideas can be understood only once others are in place. The training received by one mind fits it to grasp and expand upon ideas which gain no foothold of comprehension in another. The processes of formal education, indeed, are geared to take young (and not-so-young) minds along a genuine intellectual journey, which may even begin with ideas which are now known to be incorrect but which alone seem able to prime the system to later appreciate finer-grained truth. Such mundane facts reflect cognitive path dependence—you can't get everywhere from anywhere, and where you are now strongly constrains your potential future intellectual trajectories. In fact, such path dependence is nicely explained by treating intellectual progress as involving something like a process of computational search in a large and complex space. Previous learning inclines the system to try out certain locations in the space and not others. When the prior learning is appropriate, the job of learning some new regularity is made tractable: the prior learning acts as a filter on the space of options to be explored. Artificial neural networks that employ gradient-descent learning (see chapter 3) are especially highly constrained insofar as the learning routine forces the network always to explore at the edges of its current weight assignments. Since these constitute its current knowledge, it means that such networks cannot "jump around" in hypothesis space. The network's current location in weight space (its current knowledge) is thus a major constraint on what new "ideas" it can next explore (Elman 1994, p. 94).

In confronting devices that exhibit some degree of path dependence, the mundane observation that language allows ideas to be packaged and to migrate between individuals takes on a new force. We can now appreciate how such migrations may allow the communal construction of extremely delicate and difficult intellectual trajectories and progressions. An idea that only Joe's prior experience could make available, but that can flourish only in the intellectual niche currently provided by the brain of Mary, can now realize its full potential by journeying between Joe and

Mary as and when required. The path to a good idea can now criss-cross individual learning histories so that one agent's local minimum becomes another's potent building block. Moreover, the sheer number of intellectual niches available within a linguistically linked community provides a stunning matrix of possible inter-agent trajectories. The observation that public language allows human cognition to be collective (Churchland 1995, p. 270) thus takes on new depth once we recognize the role of such collective endeavor in transcending the path-dependent nature of individual human cognition. Even a blind and unintelligent search for productive recodings of stored data will now and again yield a powerful result. By allowing such results to migrate between individuals, culturally scaffolded reason is able to incrementally explore spaces which path-dependent individual reason could never hope to penetrate. (For a detailed, statistically based investigation of this claim, see Clark and Thornton, to appear.)

This general picture fits neatly with Merlin Donald's (1991) exploratory work on the evolution of culture and cognition. Donald recognizes very clearly the crucial role of forms of external scaffolding (particularly, of external memory systems) in human thought. But he distinguishes two major types of scaffolding, which he terms the *mythic* and the *theoretic*. Before the Greeks, Donald claims, various external formalisms were in use but were deployed only in the service of myths and narratives. The key innovation of the Greeks was to begin to use the written medium to record the *processes* of thought and argument. Whereas previous written records contained only myths or finished theories (which were to be learned wholesale and passed down relatively unaltered), the Greeks began to record partial ideas, speculations with evidence for and against them, and the like. This new practice allowed partial solutions and conjectures to be passed around, amended, completed by others, and so on. According to Donald (ibid., p. 343), what was thus created was "much more than a symbolic invention, like the alphabet, or a specific external memory medium, such as improved paper or printing"; it was "the *process* of externally encoded cognitive change and discovery."

To complete our initial inventory of the cognitive virtues of linguistically scaffolded thought, consider the physical properties of certain external media. As I construct this chapter, for example, I am continually creating, putting aside, and reorganizing chunks of text. I have files (both paper and on-line) which contain all kinds of hints and fragments, stored up over a long period of time, which may be germane to the discussion. I have source texts and papers full of notes and annotations. As I (literally, physically) move these things about, interacting first with one and then another and making new notes, annotations, and plans, the intellectual shape of the chapter grows and solidifies. It is a shape that does not spring fully developed from inner cogitations. Instead, it is the product of a sustained and iterated sequence of interactions between my brain and a variety of external props. In these cases, I am willing to say, a good deal of actual thinking involves loops and circuits that run outside the head and through the local environment. Extended intellectual arguments and theses are almost always the products of brains acting in concert with multiple external resources. These resources enable us to pursue manipulations and juxtapositions of ideas and data that would quickly baffle the un-augmented brain.[10] In all these cases, the real physical environment of printed words and symbols allows us to search, store, sequence, and reorganize data in ways alien to the onboard repertoire of the biological brain.[11]

The moral is clear. Public speech, inner rehearsal, and the use of written and on-line texts are all potent tools that reconfigure the shape of computational space. Again and again we trade culturally achieved representation against individual computation. Again and again we use words to focus, clarify, transform, offload, and control our own thinkings. Thus understood, language is not the mere imperfect mirror of our intuitive knowledge.[12] Rather, it is part and parcel of the mechanism of reason itself.

10.4 Thoughts about Thoughts: The Mangrove Effect

If a tree is seen growing on an island, which do you suppose came first? It is natural (and usually correct) to assume that the island provided the fertile soil in which a lucky seed came to rest. Mangrove forests,[13] however, constitute a revealing exception to this general rule. The mangrove grows from a floating seed which establishes itself in the water, rooting in shallow mud flats. The seedling sends complex vertical roots through the surface of the water, culminating in what looks to all intents and purposes like a small tree posing on stilts. The complex system of aerial roots, however, soon traps floating soil, weeds, and debris. After a time, the accumulation of trapped matter forms a small island. As more time passes, the island grows larger and larger. A growing mass of such islands can eventually merge, effectively extending the shoreline out to the trees. Throughout this process, and despite our prior intuitions, it is the land that is progressively built by the trees.

Something like the "mangrove effect," I suspect, is operative in some species of human thought. It is natural to suppose that words are always rooted in the fertile soil of preexisting thoughts. But sometimes, at least, the influence seems to run in the other direction. A simple example is poetry. In constructing a poem, we do not simply use words to express thoughts. Rather, it is often the properties of the words (their structure and cadence) that determine the thoughts that the poem comes to express. A similar partial reversal can occur during the construction of complex texts and arguments. By writing down our ideas, we generate a trace in a format that opens up a range of new possibilities. We can then inspect and reinspect the same ideas, coming at them from many different angles and in many different frames of mind. We can hold the original ideas steady so that we may judge them, and safely experiment with subtle alterations. We can store them in ways that allow us to compare and combine them with other complexes of ideas in ways that would quickly defeat the unaugmented imagination. In these ways, and as was remarked in the previous section, the real properties of physical text transform the space of possible thoughts.

Such observations lead me to the following conjecture: Perhaps it is public language that is responsible for a complex of rather distinctive features of human thought—viz., the ability to display *second-order cognitive dynamics*. By second-order cognitive dynamics I mean a cluster of powerful capacities involving self-evaluation, self-criticism, and finely honed remedial responses.[14] Examples would include recognizing a flaw in our own plan or argument and dedicating further cognitive efforts to fixing it, reflecting on the unreliability of our own initial judgements in certain types of situations and proceeding with special caution as a result, coming to see why we reached a particular conclusion by appreciating the logical transitions in our own thought, and thinking about the conditions under which we think best and trying to bring them about. The list could be continued, but the patten should be clear. In all these cases, we are effectively thinking about our own cognitive profiles or about specific thoughts. This "thinking about thinking" is a good candidate for a distinctively human capacity—one not evidently shared by the non-language-using animals that share our planet. Thus, it is natural to wonder whether this might be an entire species of thought in which language plays the generative role—a species of thought that is not just reflected in (or extended by) our use of words but is directly dependent upon language for its very existence. Public language and the inner rehearsal of sentences would, on this model, act like the aerial roots of the mangrove tree—the words would serves as fixed points capable of attracting and positioning additional intellectual matter, creating the islands of second-order thought so characteristic of the cognitive landscape of *Homo sapiens*.

It is easy to see, in broad outline, how this might come about. As soon as we formulate a thought in words (or on paper), it becomes an object for ourselves and for others. As an object, it is the kind of thing we can have thoughts about. In creating the object, we need have no thoughts about thoughts—but once it is there, the opportunity immediately exists to attend to it as an object in its own right. The process of linguistic formulation thus creates the stable structure to which subsequent thinkings attach.

Just such a twist on potential role of the inner rehearsal of sentences has been suggested by the linguist Ray Jackendoff. Jackendoff (to appear) suggests that the mental rehearsal of sentences may be the primary means by which our own thoughts are able to become objects of further attention and reflection. The key claim is that linguistic formulation makes complex thoughts available to processes of mental attention, and that this, in turn, opens them up to a range of further mental operations. It enables us, for example, to pick out different elements of complex thoughts and to scrutinize each in turn. It enables us to "stabilize" very abstract ideas in working memory. And it enables us to inspect and criticize our own reasoning in ways that no other representational modality allows.

What fits internal sentence-based rehearsal to play such an unusual role? The answer, I suggest, must lie in the more mundane (and temporally antecedent) role of language as an instrument of communication. In order to function as an efficient instrument of communication, public language will have been molded into a code well suited to the kinds of interpersonal exchange in which ideas are presented, inspected, and subsequently criticized. And this, in turn, involves the development of a type of code that minimizes contextuality (most words retain essentially the same meanings in the different sentences in which they occur), is effectively modality-neutral (an idea may be prompted by visual, auditory, or tactile input and yet be preserved using the same verbal formula), and allows easy rote memorization of simple strings.[15] By "freezing" our own thoughts in the memorable, context-resistant, modality-transcending format of a sentence, we thus create a special kind of mental object—an object that is amenable to scrutiny from multiple cognitive angles, is not doomed to alter or change every time we are exposed to new inputs or information, and fixes the ideas at a high level of abstraction from the idiosyncratic details of their proximal origins in sensory input. Such a mental object is, I suggest, ideally suited to figure in the evaluative, critical, and tightly focused operations distinctive of second-order cognition. It is an object fit for the close and repeated inspections highlighted by Jackendoff under the rubric of attending to our own thoughts. The coding system of public language is thus especially apt to be coopted for more private purposes of inner display, self-inspection, and self-criticism, exactly as predicted by the Vygotskian treatments mentioned in section 10.2 above. Language stands revealed as a key resource by which we effectively redescribe[16] our own thoughts in a format that makes them available for a variety of new operations and manipulations.

The emergence of such second-order cognitive dynamics is plausibly seen as one root of the veritable explosion of types and varieties of external scaffolding structures in human cultural evolution. It is because we can think about our own thinking that we can actively structure our world in ways designed to promote, support, and extend our own cognitive achievements. This process also feeds itself, as when the arrival of written text and notation allowed us to begin to fix ever more complex and extended sequences of thought and reason as objects for further scrutiny and attention. (Recall Merlin Donald's conjectures from the preceding section.)

Once the apparatus (internal and external) of sentential and text-based reflection is in place, we may expect the development of new types of non-linguistic thought and encoding—types dedicated to managing and interacting with the sentences and texts in more powerful and efficient ways.[17] The linguistic constructions, thus viewed, are a new class of objects which invite us to develop new (non-language-based) skills of use, recognition, and manipulation. Sentential and nonsentential modes of thought thus coevolve so as to complement, but not replicate, each other's special cognitive virtues.

It is a failure to appreciate this deep complementarity that, I suspect, leads Paul Churchland (one of the best and most imaginative neurophilosophers around) to dismiss linguaform expression as just a shallow reflection of our "real" knowledge. Churchland fears that without such marginalization we might mistakenly depict all thought and cognition as involving the unconscious rehearsal of sentence-like symbol strings, and thus be blinded to the powerful pattern-and-prototype-based encodings that appear to be biologically and evolutionarily fundamental. But we have now scouted much fertile intermediate territory.[18] In combining an array of biologically basic pattern-recognition skills with the special "cognitive fixatives" of word and text, we (like the mangroves) create new landscapes—new fixed points in the sea of thought. Viewed as a complementary cognitive artifact, language can genuinely extend our cognitive horizons—and without the impossible burden of recapitulating the detailed contents of nonlinguistic thought.

10.5 The Fit of Language to Brain

Consider an ill-designed artifact—for example, an early word-processing program that required extraordinary efforts to learn and was clumsy and frustrating to use. An imaginary mutant prodigy who found such a program easy would surely have needed neural resources especially pre-tuned to promote the speedy acquisition of such competence!

Now consider a superbly designed artifact: the paper clip.[19] The person who shows great speed and skill at learning to use paper clips need not be a mutant with a specially tuned brain, for the paper clip is *itself* adapted so as to facilitate easy use by beings like us (but not by rats or pigeons) in our office environment.

Suppose (just suppose) that language is like that. That is, it is an artifact that has in part evolved so as to be easily acquired and used by beings like us. It may, for instance, exhibit types of phonetic or grammatical structure that exploit particular natural biases of the human brain and perceptual system. If that were the case, it would look for all the world as if our brains were especially adapted to acquire natural language, but in fact it would be natural language that was especially adapted so as to be acquired by us, cognitive warts and all.

No doubt the truth lies somewhere in between. Recent conjectures by cognitive scientists (see e.g. Newport 1990) do suggest that certain aspects of natural languages (such as morphological structure) may be geared to exploiting windowing effects provided by the specific limitations of memory and attention found in young humans. And Christiansen (1994) has explicitly argued, from the standpoint of connectionist research, that language acquisition is empowered by a kind of symbiotic relationship between the users and the language, such that a language can persist and prosper only if it is easily learned and used by its human hosts. This symbiotic relationship forces languages to change and adapt in ways that promote learning.

Such reverse adaptation, in which natural language is to some extent adapted to the human brain, may be important in assessing the extent to which our capacity to learn and to use public language should *itself* be taken as evidence that we are cognitively very dissimilar to other animals. For humans *are*, it seems, the only animals capable of acquiring and fully exploiting the complex, abstract, open-ended symbol systems of public language.[20] Nonetheless, we need not suppose that this requires major and sweeping computational and neurological differences between us and other animals.[21] Instead, relatively minor neural changes may have made basic language learning possible for our ancestors, with the process of reverse adaptation thereafter leading to linguistic forms that more fully exploit pre-existing, language-independent cognitive biases (especially those of young humans).[22] The human brain, on this model, need not differ profoundly from the brains of higher animals. Instead, normal humans benefit from some small neurological innovation that, paired with the fantastically empowering environment of increasingly reverse-adapted public language, led to the cognitive explosions of human science, culture, and learning.

The vague and suggestive notion of reverse adaptation can even be given some (admittedly simplistic) quantitative and computational flesh. Hare and Elman (1995) used a "cultural phylogeny" of connectionist networks to model, in some detail, the series of changes that characterized the progression from the past-tense system of Old English (circa 870) to the modern system. They showed that the historical progression can be modeled, in some detail, by a series of neural networks in which the output from one generation is used as the training data for the next. This process yields changes in the language itself as the language alters to reflect the learning profiles of its users. Briefly, this is what happens: An original network is trained on the Old English forms. A second network is then trained (though not to perfection) on the forms produced by the first. This output is then used to train a further network, and so on. Crucially, any errors one network makes in learning to perform the mappings become parts of the next network's data set. Patterns that are hard to learn and items that are close in form to other, differently inflected items tend to disappear. As Hare and Elman (ibid., p. 61) put it: "At the onset, the classes [of verbs] differ in terms of their phonological coherence and their class size. Those patterns that are initially less common or less well defined are the hardest to learn. And these tend to be lost over several generations of learning. This process snowballs as the dominant class gathers in new members and this combined class becomes an ever more powerful attractor." By thus studying the interplay between the external data set and the processes of individual learning, Hare and Elman were able to make some quite fine-grained predictions (borne out by the linguistic facts) about the historical progression from Old English to Modern English. The important moral, for our purposes, is that in such cases the external scaffoldings of cognition *themselves* adapt so as to better prosper in the niche provided by human brains. The complementarity between the biological brain and its artifactual props and supports is thus enforced by coevolutionary forces uniting user and artifact in a virtuous circle of mutual modulation.

10.6 Where Does the Mind Stop and the Rest of the World Begin?[23]

The complexities of user-artifact dynamics invite reflection on a more general topic: how to conceive the boundary between the intelligent system and the world. This boundary, as we saw in previous chapters, looks to be rather more plastic than had previously been supposed—in many cases, selected extra-bodily resources constitute important parts of extended computational and cognitive processes. Taken to extremes, this seepage of the mind into the world threatens to reconfigure our fundamental self-image by broadening our view of persons to include, at times, aspects of the local environment. This kind of broadening is probably most plausible in cases involving the external props of written text and spoken words, for interactions with these external media are ubiquitous (in educated modern cultures), reliable, and developmentally basic. Human brains, in such cultures, come to expect the surrounding media of text and speech as surely as they expect to function in a world of weight, force, friction, and gravity. Language is a constant, and as such it can be safely relied upon as the backdrop against which on-line processes of neural computation develop. Just as a neural-network controller for moving an arm to a target in space will define its commands to factor in the spring of muscles and the effects of gravity, so the processes of onboard reason may learn to factor in the potential contributions of textual offloading and reorganization, and vocal rehearsal and exchange. The mature cognitive competencies which we identify as mind and intellect may thus be more like ship navigation (see chapter 3) than capacities of the bare biological brain. Ship navigation emerges from the well-orchestrated adaptation of an extended complex system comprising individuals, instruments, and practices. Much of what we commonly identify as our mental capacities may likewise, I suspect, turn out to be properties of the wider, environmentally extended systems of which human brains are just one (important) part.

This is a big claim, and I do not expect to convince the skeptics here. But it is not, I think, quite as wild as it may at first appear. There is, after all, a quite general difficulty in drawing a firm line between a user and a tool.[24] A stone held in one's hand and used to crack a nut is clearly a tool. But if a bird drops a nut from the air so that it will break on contact with the ground, is the ground a tool? Some birds swallow small stones to aid digestion—are the stones tools, or, once ingested, simply parts of the bird? Is a tree, once climbed to escape a predator, a tool? What about a spider's web?

Public language and the props of text and symbolic notation are, I suggest, not unlike the stones swallowed by birds. The question "Where does the user end and the tool begin?" invites, in both cases, a delicate call. In the light of the larger body of our previous discussions, I am at a minimum persuaded of two claims. The first is that some human actions are more like thoughts than they at first appear. These are the actions whose true goal is to alter the computational tasks that confront the brain as we try to solve a problem—what Kirsh and Maglio called "epistemic actions." The second is that certain harms to the environment may have the kind of moral significance we normally associate with harm to the person—I am thinking here especially of the cases, described in chapter 3 above, of neurologically impaired humans who get along by adding especially dense layers of external prompts and supports to their daily surroundings. Tampering with these supports, it seems to me, would be more akin to a crime against the person than to a crime against property. In a similar vein, Clark and Chalmers (1995) describe the case of a neurologically impaired agent who relies heavily on a constantly carried notebook deferring to its contents on numerous daily occasions. Wanton destruction of the notebook, in such a case, has an especially worrying moral aspect: it is surely harm to the person, in about as literal a sense as can be imagined.

In the light of these concerns and the apparent methodological value (see chapters 3, 4, 6, and 8 above) of studying extended brain-body-world systems as integrated computational and dynamic wholes, I am convinced that it is valuable to (at times) treat cognitive processes as extending beyond the narrow confines of skin and skull. And I am led to wonder whether the intuitive notion of mind itself should not be broadened so as to encompass a variety of external props and aids—whether, that is, the system we often refer to as "mind" is in fact much wider than the one we call "brain." Such a more general conclusion may at first seem unpalatable. One reason, I think, is that we are prone to confuse the mental with the conscious. And I assuredly do not seek to claim that individual consciousness extends outside the head. It seems clear, however, that not everything that occurs in the brain and constitutes (in current scientific usage) a mental or cognitive process is tied up with *conscious* processing.[25] More plausibly, it may be suggested that what keeps real mental and cognitive processes in the head is some consideration of portability. That is to say, we are moved by a vision of what might be called the Naked

Mind: a vision of the resources and operations we can *always* bring to bear on a cognitive task, regardless of whatever further opportunities the local environment may or may not afford us.

I am sympathetic to this objection. It seems clear that the brain (or perhaps, on this view, the brain and body) is a proper and distinct object of study and interest. And what makes it such is precisely the fact that it comprises some such set of core, basic, portable cognitive resources. These resources may incorporate bodily actions as integral parts of some cognitive processes (as when we use our fingers to offload working memory in the context of a tricky calculation). But they will not encompass the more contingent aspects of our external environment—the ones that may come and go, such as a pocket calculator. Nonetheless, I do not think that the portability consideration can ultimately bear sufficient conceptual weight, and for two reasons. First, there is a risk of begging the question. If we ask *why* portability should matter to the constitution of specific mental or cognitive processes, the only answer seems to be that we want such processes to come in a distinct, individually mobile package. But this, of course, is just to invoke the boundary of skin and/or skull all over again—and it is the legitimacy of this very boundary that is in question. Second, it would be easy (albeit a little tedious for the reader) to construct a variety of troublesome cases. What if some people *always* carried a pocket calculator; what if we one day have such devices implanted in our brains? What if we have "body docks" for a variety of such devices and "dress" each day by adding on devices appropriate for that day's prescribed problem-solving activity? Nor can the vulnerability of such additional devices to discrete damage or malfunction serve to distinguish them, for the biological brain likewise is at risk of losing specific problem-solving capacities through lesion or trauma.

The most compelling source of our anxieties, however, probably concerns that most puzzling entity, the *self*.[26] Does the putative spread of mental and cognitive processes out into the world imply some correlative (and surely unsettling) leakage of the self into the local surroundings? The answer now looks to be (sorry!) "Yes and No." No, because (as has already been conceded) conscious contents supervene on individual brains. But Yes, because such conscious episodes are at best snapshots of the self considered as an evolving psychological profile. Thoughts, considered only as snapshots of our conscious mental activity, are fully explained, I am willing to say, by the current state of the brain. But the flow of reason and thoughts, and the temporal evolution of ideas and attitudes, are determined and explained by the intimate, complex, continued interplay of brain, body, and world. It is, if you like, a genuine aspect of my psychological profile to be the kind of person who writes a book like this—despite the fact that the flow and shape of the ideas expressed depended profoundly on a variety of iterated interactions between my biological brain and a small army of external encodings, recodings, and structuring resources.

Such liberality about cognitive processes and cognitive profiles must, of course, be balanced by a good helping of common sense. Mind cannot usefully be extended willy-nilly into the world. There would be little value in an analysis that credited me with knowing all the facts in the *Encyclopaedia Britannica* just because I paid the monthly installments and found space for it in my garage. Nor should the distinction between my mind and yours be allowed to collapse just because we are found chatting on the bus. What, then, distinguishes the more plausible cases of robust cognitive extension from the rest?

Some important features of the more plausible cases (such as the neurologically impaired agent's notebook) can be isolated quickly. The notebook is always there—it is not locked in the garage, or rarely consulted. The information it contains is easy to access and use. The information is automatically endorsed—not subject to critical scrutiny, unlike the musings of a companion on a bus. Finally, the information was originally gathered and endorsed by the current user (unlike the entries in the encyclopedia). These conditions may not all be essential. And there may be others I have missed. But the overall picture is of a rather special kind of user/artifact relationship—one in which the artifact is reliable present, frequently used, personally "tailored," and deeply trusted. Human agents, as we saw on numerous occasions in previous chapters, may press all kinds of crucial cognitive and computational benefits from interactions with artifacts that lack one or all of these features. But it is probably only when something like these conditions are met that we can plausibly argue for an extension of the morally resonant notions of self, mind, and agenthood to include aspects of the world beyond the skin. It is thus only when the relationship between user and artifact is about as close and intimate as that between the spider and the web[27] that the bounds of the self—and not just those of computation and broadly cognitive process—threaten to push out into the world.

The crucial point in the case of the agent and the notebook is that the entries in the notebook play the same explanatory role,[28] with respect to the agent's behavior, as would a piece of information encoded in long-term memory. The special conditions (accessibility, automatic endorsement, etc.) are necessary to ensure this kind of functional isomorphism. However, even if one grants (as many will not) that such an isomorphism obtains, it may be possible to avoid the radical conclusion concerning distributed agenthood. An alternative (and, I think, equally acceptable) conclusion would be that the agent remains locked within the envelope of skin and skull, but that beliefs, knowledge, and perhaps other mental states now depend on physical vehicles that can (at times) spread out to include select aspects of the local environment. Such a picture preserves the idea of the agent as the combination of body and biological brain, and hence allows us to speak—as we surely should—of the agent's sometimes manipulating and structuring those same external resources in ways designed to further extend, offload, or transform her own basic problem-solving activities. But it allows also that in this "reaching out" to the world we sometimes create wider cognitive and computational webs: webs whose understanding and analysis requires the application of the tools and concepts of cognitive science to larger, hybrid entities comprising brains, bodies, and a wide variety of external structures and processes.

In sum, I am content to let the notions of self and agency fall where they will. In the final analysis, I assert only that we have, at a minimum, good explanatory and methodological reasons to (at times) embrace a quite liberal notion of the scope of computation and cognitive processes—one that explicitly allows the spread of such processes across brain, body, world, and artifact. Paramount among such artifacts are the various manifestations of public language. Language is in many ways the ultimate artifact: so ubiquitous it is almost invisible, so intimate it is not clear whether it is a kind of tool or a dimension of the user. Whatever the boundaries, we confront at the very least a tightly linked economy in which the biological brain is fantastically empowered by some of its strangest and most recent creations: words in the air, symbols on the printed page.

11

Minds, Brains, and Tuna: A Summary in Brine

The swimming capacities of many fishes, such as dolphins and bluefin tuna, are staggering. These aquatic beings far outperform anything that nautical science has so far produced. Such fish are both mavericks of maneuverability and, it seems, paradoxes of propulsion. It is estimated that the dolphin, for example, is simply not strong enough[1] to propel itself at the speeds it is observed to reach. In attempting to unravel this mystery, two experts in fluid dynamics, the brothers Michael and George Triantafyllou, have been led to an interesting hypothesis: that the extraordinary swimming efficiency of certain fishes is due to an evolved capacity to exploit and create additional sources of kinetic energy in the watery environment. Such fishes, it seems, exploit aquatic swirls, eddies, and vortices to "turbocharge" propulsion and aid maneuverability. Such fluid phenomena sometimes occur naturally (e.g., where flowing water hits a rock). But the fish's exploitation of such external aids does not stop there. Instead, the fish actively creates a variety of vortices and pressure gradients (e.g. by flapping its tail) and then uses these to support subsequent speedy, agile behavior. By thus controlling and exploiting local environmental structure, the fish is able to produce fast starts and turns that make our ocean-going vessels look clumsy, ponderous, and laggardly. "Aided by a continuous parade of such vortices," Triantafyllou and Triantafyllou (1995, p. 69) point out, "it is even possible for a fish's swimming efficiency to exceed 100 percent." Ships and submarines reap no such benefits: they treat the aquatic environment as an obstacle to be negotiated and do not seek to subvert it to their own ends by monitoring and massaging the fluid dynamics surrounding the hull.

The tale of the tuna[2] reminds us that biological systems profit *profoundly* from local environmental structure. The environment is not best conceived solely as a problem domain to be negotiated. It is equally, and crucially, a resource to be factored into the solutions. This simple observation has, as we have seen, some far-reaching consequences.

First and foremost, we must recognize the brain for what it is. Ours are not the brains of disembodied spirits conveniently glued into ambulant, corporeal shells of flesh and blood. Rather, they are *essentially* the brains of embodied agents capable of creating and exploiting structure in the world. Conceived as controllers of embodied action, brains will sometimes devote considerable energy not to the direct, one-stop solution of a problem, but to the control and exploitation of environmental structures. Such structures, molded by an iterated sequence of brain-world interactions, can alter and transform the original problem until it takes a form that can be managed with the limited resources of pattern-completing, neural-network-style cognition.

Second, we should therefore beware of mistaking the problem-solving profile of the embodied, socially and environmentally embedded mind for that of the basic brain. Just because humans can do logic and science, we should not assume that the brain contains a full-blown logic engine or that it encodes scientific theories in ways akin to their standard expression in words and sentences.[3] Instead, both logic and science rely heavily on the use and manipulation of external media, especially the formalisms of language and logic and the capacities of storage, transmission, and refinement provided by cultural institutions and by the use of spoken and written text. These resources, I have argued, are best seen as alien but *complementary* to the brain's style of storage and computation. The brain need not waste its time *replicating* such capacities. Rather, it must learn to interface[4] with the external media in ways that maximally exploit their peculiar virtues.

Third, we must begin to face up to some rather puzzling (dare I say metaphysical?) questions. For starters, the nature and the bounds of the intelligent agent look increasingly fuzzy. Gone is the central executive[5] in the brain—the real boss who organizes and integrates the activities of multiple special-purpose subsystems. And gone is the neat boundary between the thinker (the bodiless intellectual engine) and the thinker's world. In place of this comforting image we confront a vision of mind as a grab bag of inner agencies whose computational roles are often best described by including aspects of the local environment (both in complex control loops and in a wide variety of informational transformations and manipulations). In light of all this, it may for some purposes be wise to consider the intelligent system as a spatio-temporally extended process not limited by the tenuous envelope of skin and skull.[6] Less dramatically, the traditional divisions among perception, cognition, and action[7] look increasingly unhelpful. With the demise of the central executive, perception and cognition look harder to distinguish in the brain. And the division between thought and action fragments once we recognize that real-world actions often play precisely the kinds of functional roles more usually associated with internal processes of cognition and computation.

Fourth (and last), whatever the metaphysical niceties, there are immediate and pressing methodological morals. Cognitive science, if the embodied, embedded perspective is even halfway on target, can no longer afford the individualistic, isolationist biases that characterized its early decades. We now need a wider view—one that incorporates a multiplicity of ecological and cultural approaches as well as the traditional core of neuroscience, linguistics, and artificial intelligence. And we need new tools with which to investigate effects that span multiple time scales, involve multiple individuals, and incorporate complex environmental interactions. A canny combination of Dynamical Systems approaches, real-world robotics, and large-scale simulations (of evolutionary and collective effects) is, at present, probably the best we can do. But such investigations, I argued, must be carefully interlocked with real ongoing neuroscientific research, and thus anchored, whenever possible, in knowledge about the biological brain. In pursuit of this interlock, it would be folly to simply jettison the hard-won bedrock of cognitive scientific understanding that involves ideas of internal representation and computation. The true lesson of our investigations of embodied, embedded cognition is not that we somehow succeed *without* representing (or, worse, without computing). Rather, it is that the *kinds* of internal representation and computation we employ are selected so as to complement the complex social and ecological settings in which we must act. Thus, we ignore or downplay such wider settings at our intellectual peril.

And there it is. The end of a long and surely unfinished journey. There were loops, detours, and, to be sure, one or two roadblocks more circumnavigated than demolished. Much remains to be done. I hope I've pulled together some threads, built a few bridges, and highlighted some pressing issues. Like Humpty Dumpty, brain, body, and world are going to take a *whole lot* of putting back together again. But it's worth persevering because until these parts click into place we will never see ourselves aright or appreciate the complex conspiracy that is adaptive success.

Epilogue: A Brain Speaks[1]

I am John's brain.[2] In the flesh, I am just a rather undistinguished looking gray-white mass of cells. My surface is heavily convoluted, and I am possessed of a fairly differentiated internal structure. John and I are on rather close and intimate terms; indeed, sometimes it is hard to tell us apart. But at times John takes this intimacy a little too far. When that happens, he gets very confused about my role and my functioning. He imagines that I organize and process information in ways that echo his own perspective on the world. In short, he thinks that his thoughts are, in a rather direct sense, my thoughts. There is some truth to this, of course. But things are really rather more complicated than John suspects, as I shall try to show.

In the first place, John is congenitally blind to the bulk of my daily activities. At best, he catches occasional glimpses and distorted shadows of my real work. Generally speaking, these fleeting glimpses portray only the products of my vast subterranean activity, rather than the processes that give rise to them. Such products include the play of mental images and the steps in a logical train of thought or flow of ideas.

Moreover, John's access to these products is a pretty rough and ready affair. What filters into his conscious awareness is somewhat akin to what gets onto the screen display of a personal computer. In both cases, what is displayed is just a specially tailored summary of the results of certain episodes of internal activity: results for which the user has some particular use. Evolution, after all, would not waste time and money (search and energy) to display to John a faithful record of inner goings on unless they could help John to hunt, survive, and reproduce. John, as a result, is apprised of only the bare minimum of knowledge about my inner activities. All he needs to know is the overall significance of the upshots of a select few of these activities: that part of me is in a state associated with the presence of a dangerous predator and that flight is therefore indicated, and other things of that sort. What John (the conscious agent) gets from me is thus rather like what a driver gets from an electronic dashboard display: information pertaining to the few inner and outer parameters to which his gross considered activity can make a useful difference.

A complex of important misapprehensions center around the question of the provenance of thoughts. John thinks of me as the point source of the intellectual products he identifies as his thoughts. But, to put it crudely, I do not have John's thoughts. John has John's thoughts, and I am just one item in the array of physical events and processes that enable the thinking to occur. John is an agent whose nature is fixed by a complex interplay involving a mass of internal goings on (including my activity), a particular kind of physical embodiment, and a certain embedding in the world. The combination of embodiment and embedding provides for persistent informational and physical couplings between John and his world—couplings that leave much of John's "knowledge" out in the world and available for retrieval, transformation, and use as and when required.

Take this simple example: A few days ago, John sat at his desk and worked rather hard for a sustained period of time. Eventually he got up and left his office, satisfied with his day's work. "My brain," he reflected (for he prides himself on his physicalism), "has done very well. It has come up with some neat ideas." John's image of the events of the day depicted me as the point source of those ideas—ideas which he thinks he captured on paper as a mere convenience and a hedge against forgetting. I am, of course, grateful that John gives me so much credit. He attributes the finished intellectual products directly to me. But in this case, at least, the credit should be extended a little further. My role in the origination of these intellectual products is certainly a vital one: destroy me and the intellectual productivity will surely cease! But my role is more delicately constituted then John's simple image suggests. Those ideas of which he is so proud did not spring fully formed out of my activity. If truth be told, I acted rather as a mediating factor in some complex feedback loops encompassing John and selected chunks of his local environment. Bluntly, I spent the day in a variety of close and complex interactions with a number of external props. Without these, the finished intellectual products would never have taken shape. My role, as best I can recall, was to support John's rereading of a bunch of old materials and notes, and to react to those materials by producing a few fragmentary ideas and criticisms. These small responses were stored as further marks on paper and in margins. Later on, I played a role in the reorganization of these marks on clean sheets of paper, adding new on-line reactions to the fragmentary ideas. The cycle of reading, responding, and external reorganization was repeated again and again. At the end of the day, the "good ideas" with which John was so quick to credit me emerged as the fruits of these repeated little interactions between me and the various external media. Credit thus belongs not so much to me as to the spatially and temporally extended process in which I played a role.

On reflection, John would probably agree to this description of my role on that day. But I would caution him that even this can be misleading. So far, I have allowed myself to speak as if I were a unified inner resource contributing to these interactive episodes. This is an illusion which the present literary device encourages and which John seems to share. But once again, if truth be told, I am not one inner voice but many. I am so many inner voices, in fact, that the metaphor of the inner voice must itself mislead, for it surely suggests inner subagencies of some sophistication and perhaps possessing a rudimentary self-consciousness. In reality, I consist only of multiple mindless streams of highly parallel and often relatively independent computational processes. I am not a mass of little agents so much as a mass of non-agents, tuned and responsive to proprietary inputs and cleverly orchestrated by evolution so as to yield successful purposive behavior in most daily settings. My single voice, then, is no more than a literary conceit.

At root, John's mistakes are all variations on a single theme. He thinks that I see the world as he does, that I parcel things up as he would, and that I think the way he would report his thoughts. None of this is the case. I am not the inner echo of John's conceptualizations. Rather, I am their somewhat alien source. To see just how alien I can be, John need only reflect on some of the rather extraordinary and unexpected ways that damage to me (the brain) can affect the cognitive profiles of beings like John. Damage to me could, for example, result in the selective impairment of John's capacity to recall the names of small manipulable objects yet leave unscathed his capacity to name larger ones. The reason for this has to do with my storing and retrieving heavily visually oriented information in ways distinct from those I deploy for heavily functionally oriented information; the former mode helps pick out the large items and the latter the small ones. The point is that this facet of my internal organization is altogether alien to John—it respects needs, principles, and opportunities of which John is blissfully unaware. Unfortunately, instead of trying to comprehend my modes of information storage in their own terms, John prefers to imagine that I organize my knowledge the way he—heavily influenced by the particular words in his language—organizes his. Thus, he supposes that I store information in clusters that respect what he calls "concepts" (generally, names that figure in his linguistic classifications of worldly events, states, and processes). Here, as usual, John is far too quick to identify my organization with his own perspective. Certainly I store and access bodies of information—bodies which together, if I am functioning normally, support a wide range of successful uses of words and a variety of interactions with the physical and social worlds. But the "concepts" that so occupy John's imagination correspond only to public names for grab bags of knowledge and abilities whose neural underpinnings are in fact many and various. John's "concepts" do not correspond to anything especially unified, as far as I am concerned. And why should they? The situation is rather like that of a person who can build a boat. To speak of the ability to build a boat is to use a simple phrase to ascribe a panoply of skills whose cognitive and physical underpinnings vary greatly. The unity exists only insofar as that particular grab bag of cognitive and physical skills has special significance for a community of seafaring agents. John's "concepts," it seems to me, are just like that: names for complexes of skills whose unity rests not on facts about me but on facts about John's way of life.

John's tendency to hallucinate his own perspective onto me extends to his conception of my knowledge of the external world. John walks around and feels as if he commands a stable three-dimensional image of his immediate surroundings. John's feelings notwithstanding, I command no such thing. I register small regions of detail in rapid succession as I fixate first on this and then on that aspect of the visual scene. And I do not trouble myself to store all that detail in some internal model that requires constant maintenance and updating. Instead, I am adept at revisiting parts of the scene so as to re-create detailed knowledge as and when required. As a result of this trick, and others, John has such a fluent capacity to negotiate his local environment that he thinks he commands a constant inner vision of the details of his surroundings. In truth, what John sees has more to do with the abilities I confer on him to interact constantly, in real time, with rich external sources of information than with the kind of passive and enduring registration of information in terms of which he conceives his own seeings.

The sad fact, then, is that almost nothing about me is the way John imagines it to be. We remain strangers despite our intimacy (or perhaps because of it). John's language, introspections, and oversimplistic physicalism incline him to identify my organization too closely with his own limited perspective. He is thus blind to my fragmentary, opportunistic, and generally alien nature. He forgets that I am in large part a survival-oriented device that greatly predates the emergence of linguistic abilities, and that my role in promoting conscious and linguaform cognition is just a recent sideline. This sideline is, or course, a major root of his misconceptions. Possessed as John is of such a magnificent vehicle for the compact and communicable expression and manipulation of knowledge, he often mistakes the forms and conventions of that linguistic vehicle for the structure of neural activity itself.

But hope springs eternal (more or less). I am of late heartened by the emergence of new investigative techniques, such as non-invasive brain imaging, the study of artificial neural networks, and research in real-world robotics. Such studies and techniques bode well for a better understanding of the very complex relations among my activity, the local environment, and the patchwork construction of the sense of self. In the meantime, just bear in mind that, despite our intimacy, John really knows very little about me. Think of me as the Martian in John's head.[3]

Notes

Preface

1. Descartes depicted mind as an immaterial substance that communicated with the body via the interface of the pineal gland. See, e.g., Meditations II and IV in *The Philosophical Works of Descartes* (Cambridge University Press, 1991).

2. See Gilbert Ryle, *The Concept of Mind* (Hutchinson, 1949).

3. AI is the study of how to get computers to perform tasks that might be described as requiring intelligence, knowledge, or understanding.

4. Sloman, "Notes on consciousness," *AISB Quarterly* 72 (1990): 8–14.

5. Deep Thought is a chess program that plays at the grandmaster level. It relies on extensive search, examining around a billion possible moves per second. Human chess experts, by contrast, seem to use less search and to rely on very different styles of reasoning—see, e.g., H. Simon and K. Gilmartin, "A simulation of memory for chess positions," *Cognitive Psychology* 5 (1973): 29–46.

6. The example is cited in Michie and Johnson 1984 and reported in Clark 1989. The quoted passage is from p. 95 of Michie and Johnson.

7. The clearest existing expression of such a view is probably the 'enaction framework' developed in Varela et al. 1991.

8. See especially Dennett 1991.

Chapter 1

1. The material concerning Dante II is based on a report by Peter Monaghen (*Chronicle of Higher Education*, August 10, 1994, pp. A6–A8).

2. See e.g. W. Grey Walter, "An imitation of life," *Scientific American* 182 (1959), no. 5: 42–45 ; Steven Levy, *Artificial Life: The Quest for a New Creation* (Pantheon, 1992), pp. 283–284.

3. Early work in AI, such as Newell and Simon's (1972) work on the General Problem Solver, tended to stress reasoning and all-purpose problem solving. Soon, however, it became evident that for many purposes a rich and detailed knowledge

base tailored to a specific domain of activity was a crucial determinant of success. This realization led to an explosion of work in so-called expert systems, which were provided with task-specific data elicited from human experts and which were thus able to achieve quite high degrees of competence in restricted domains such as medical diagnosis. The program MYCIN (Shortliffe 1976) relied on a body of explicitly formulated rules and guidelines, such as the following rule for blood injections: "If (1) the site of culture is blood, (2) the Gram stain of the organism is gramneg, (3) the morphology of the organism rod, and (4) the patient is a compromised host, then there is suggestive evidence that the identity of the organism is *pseudomonas aeruginose*" (Feigenbaum 1977, p. 1016). Such systems proved brittle and restricted. They rapidly degenerate into automated idiocy if the user steps over a thin red line of grammar or expression or uses terms that have rich real-world significance not explicitly reflected in the task-specific data base (e.g., the rusty Chevy diagnosed as having measles; see Lenat and Feigenbaum 1992, p. 197). How can this slide into idiocy be averted? One possibility is that all that is needed is to "turbocharge" the kinds of traditional approaches mentioned above. SOAR (Laird et al. 1987) is an attempt to create a more powerful version of the General Problem Solver. CYC (see the introduction to this volume and Lenat and Feigenbaum 1992) is an attempt to create a much larger and richer knowledge base. SOAR and CYC share a commitment to the extensive use of traditional text-inspired symbolic forms of knowledge and goal encoding. But it may be that the fundamental problem lies in the traditional approach itself: that the model of intelligence as the disembodied manipulation of strings of symbols inside the head or computer is itself mistaken. The present treatment explores some of the alternatives.

4. On Herbert see Connell 1989.

5. Ron McClamrock (1995) reports a nice case from Marr in which a control loop runs outside the head and into the local environment. In McClamrock's words (p. 85): "Flies, it turns out, don't quite know that to fly they should flap their wings. They don't take off by sending some signal from the brain to the wings. Rather, there is a direct control link from the fly's feet to its wings, such that when the feet cease to be in contact with a surface, the fly's wings begin to flap. To take off, the fly simply jumps and then lets the signal from the feet trigger the wings."

6. Attila, described on pp. 300–301 of Levy's *Artificial Life*, was designed by Colin Angle and Rodney Brooks. A predecessor, Genghis, is nicely described in Brooks 1993.

7. On connectionism see chapter 4 below. Q-learning is a form of reinforcement learning (see Kaelbling 1993 and Sutton 1991) developed by Watkins (1989). The use of neural networks in Q-learning scenarios is discussed in Lin 1993.

8. It is nicely presented in Churchland et al. 1994, and it permeates much of Dennett 1991.

9. This research was conducted by Zolten Dienes at the University of Sussex (personal communication).

10. Especially Dennett 1991, Ballard 1991, and Churchland et al. 1994.

11. See Mackay 1967 and MacKay 1973. I first encountered this example in O'Regan 1992 (pp. 471–476).

12. "The 'percept' of the bottle is an *action*, namely the visual or mental exploration of the bottle. It is *not* simply the passive sensation we get from the retina or some iconic derivative of the information upon it." (O'Regan 1992, p. 472)

13. See McConkie and Rayner 1976, McConkie 1979, McConkie 1990, O'Regan 1990, and Rayner et al. 1980.

Chapter 2

1. See e.g. Piaget 1952, 1976; Gibson 1979; Bruner 1968; Vygotsky 1986.

2. For discussions of action loop phenomena see Cole et al. 1978; Rutkowska 1986, 1993; Thelen and Smith 1994.

3. See also Rutkowska 1993, p. 60.

4. The work reported was carried out by Adolph et al. (1993).

5. See Shields and Rovee-Collier 1992; Rovee-Collier 1990.

6. For a general survey of distorting-lens experiments see Welch 1978.

7. Recall the animate-vision examples described in chapter 1.

8. See e.g. Gesell 1939, McGraw 1945, and chapter 1 of Thelen and Smith 1994.

9. For a computational simulation of this, and other emergent phenomena see Resnick 1994, pp. 60–67.

10. See Thelen and Smith 1994, pp. 11–12. See also Thelen et al. 1982 and Thelen et al. 1984.

11. See also Thelen 1986, Thelen et al. 1987, and Thelen and Ulrich 1991.

12. The example is from Maes 1994 (pp. 145–146). Classical scheduling agents are described in Kleinrock and Nilsson 1981.

13. It is based on Malone et al. 1988.

14. Variability is thus data, *not* noise—see Smith and Thelen 1994, pp. 86–88.

15. Polit and Bizzi 1978; Hogan et al. 1987; Jordan et al. 1994; Thelen and Smith 1994.

16. See e.g. Vygotsky 1986.

17. See e.g. Valsiner 1987 and Wertsch 1981.

18. For this usage see e.g. Rutkowska 1993, pp. 79–80.

19. See also chapter 4 of Clark 1989 and chapter 3 of Rutkowska 1993.

20. See Vogel 1981 and chapter 4 of Clark 1989.

21. See Millikan 1995 and Clark 1995. A computation-oriented incarnation of such ideas can also be found in Rutkowska's (1993, pp. 67–78) use of "action programs" as a foundational construct in developmental theories.

Chapter 3

1. Some of the original ideas were formulated as long ago (in AI terms) as 1943—see McCulloch and Pitts 1943, Hebb 1949, and Rosenblatt 1962.

2. Consider the Mataric model, described in section 2.6 above. The kind of map Mataric details has strong affinities with recent models of how the hippocampus may encode spatial information (McNaughton 1989). One disanology, however, concerns the use in Mataric's model of single nodes as encoding landmark information. The hippocampus probably uses a much more distributed form of representation, with many neurons involved in representing each landmark. There exist more detailed artificial-neural-network-style models of hippocampal function that do indeed recognize the role of such distribution (see e.g. O'Keefe 1989 and McNaughton and Nadel 1990). Such models suggest that the hippocampal structure is a very good candidate for a real neural system that operates in ways broadly similar to the artificial neural networks described in section 3.2. It is equally evident, however, that more neurobiologically realistic models will need to incorporate many features not found in the majority of artificial networks. For example, the kind of *highly detailed* error-correcting feedback used by back-propagation learning devices is probably not found in the brain, although some kind of error-driven adaptation certainly takes place. Nor does real neural circuitry exhibit the symmetrical connectivity displayed in most artificial networks; instead, we often confront asymmetric, special-purpose connectivity. Despite such differences (and there are plenty more—see McNaughton 1989 and Churchland and Sejnowski 1992), computational models of real neural structures still owe much more to the frameworks of artificial neural networks than to those of classical AI. And the bedrock capacity responsible for this is the reliance on associative memory systems, which replace rule-and-symbol reasoning with rich and powerful processes of pattern completion.

3. Digital Equipment Corporation DTC-01-AA.

4. Such functions are usually nonlinear; i.e., the strength of the output is not directly proportional to the sum of the inputs. Instead, it may be (for example) proportional when the incoming signals are of medium intensity, but flatten out when they are very strong or very weak.

5. The response characteristics of hidden units were described above.

6. For an especially clear and accessible account of such approaches, see Churchland 1995. See also Clark 1989 and Churchland 1989.

7. See Rumelhart and McClelland 1986 and the critical assessments in Clark 1989 and Clark 1993.

8. See Elman 1991.

9. McClelland 1989; Plunkett and Sinha 1991.

10. For a much more careful delineation of this class of models, see Clark 1993.

11. Full details can be found in chapter 5 of Clark 1989.

12. This is, of course, open to question. But it seems increasingly clear that, whatever brains are really like, they are closer to the information-processing profile of artificial neural networks than to that of classical devices. In fact, it seems likely (see section 3.4 below) that biological brains exploit more special-purpose machinery than typical artificial neural networks, but that the style of representation and processing remains similar along several major dimensions (such as the use of parallel distributed encodings and vector to vector transformations—see e.g. Churchland 1989, Churchland and Sejnowski 1992, and Churchland 1995).

13. See especially the discussions of collective activity (chapter 4) and of the wider role of language and culture (chapters 9 and 10).

14. *Parallel Distributed Processing: Explorations in the Microstructure of Cognition*, volume 1: *Foundations* and volume 2: *Psychological and Biological Models* (MIT Press, 1986). The work described is found in chapter 14 (i.e. Rumelhart et al. 1986).

15. See e.g. Vygotsky 1962. See also chapters 9 and 10 of the present volume.

16. In Clark 1986 and Clark 1988a I discuss results from the domain of drawing that lend further support to such a conjecture. In these papers I also discuss work by Chambers and Reisberg (1985) on the special properties of actual drawings as opposed to mental images of drawings. This research is also cited in Kirsh 1995 and in Zhang and Norman 1994.

17. This theme is explored at length in chapter 10 below.

18. See Kirsh and Maglio's (1994, p. 515) comments concerning the need to redefine the state space in which planning occurs.

19. I thank Caroline Baum, Director of the Occupational Therapy Unit at Washington University School of Medicine, for bringing these cases to my attention. See Baum 1993 and Edwards et al. 1994.

20. See e.g. Suchman 1987 and Bratman et al. 1991.

21. This is, of course, no longer strictly true. Artificial neural networks themselves constitute such external pattern-completing resources (Churchland 1995, chapter 11). Moreover, other agents and animals also constitute pattern-completion resources external to the individual. More on this in chapter 4.

22. The epilogue illustrates this claim by considering the role of the brain in generating a complex linked stream of ideas. See also section 10.5.

Chapter 4

1. The story is based on accounts in Alexopoulos and Mims 1979 and Farr 1981.

2. There are two main types: acellular slime molds, in which the cells fuse to form a multinucleate mass, and cellular slime molds, in which the cells aggregate but never fuse and in which the aggregate multicellular mass forms a mobile body (sometimes called a *slug* or a *grex*). See chapter 1 of Ashworth and Dee 1975.

3. The account of *D. discoideum* is based on Ashworth and Dee 1975, pp. 32–36.

4. Acellular slime molds, like *Fuligo septico*, do not form a motile grex. Instead, the plasmodium migrates by a process of protoplasmic streaming.

5. Here, as elsewhere in sections 4.1 and 4.2, I follow the lead of Mitchel Resnick, whose book *Turtles, Termites, and Traffic Jams* (1994) is both a paradigm of clarity and a powerful testimony to the scope and the power of decentralized thinking.

6. 'Stigmergic', a combination of 'stigma' (sign) and 'ergon' (work), connotes the use of work as the signal for more work.

7. See Grasse 1959 and Beckers et al. 1994.

8. On many vessels there is in fact a formal plan. But crew members do not explicitly use it to structure their actions; indeed, Hutchins (1995, p. 178) suggests, the plan would not work even if they did.

9. True stigmergy requires a complete lack of flexibility of response in the presence of a triggering condition. Thus, human activity is, in general, only quasi-stigmergic. What is common is the use of environmental conditions as instigators of action and the overall ability of the group to perform a problem-solving activity that exceeds the knowledge and the computational scope of each individual member.

10. Hutchins (1995, chapter 3) describes these in detail. An alidade is a kind of telescopic sighting device; a hoey is a one-armed protractor used for marking lines on charts.

11. See p. 171 of Hutchins 1995 and chapters 3 and 10 of the present volume.

12. I don't wish (or need) to beg questions concerning the relative contributions of blind variation and natural selection, on the one hand, and more fundamental self-organizing properties of matter, chemicals, and cells, on the other. For some discussion, see Kauffman 1993, chapter 8 of Dennett 1995, and pp. 180–214 of Varela et al. 1991. Thanks to Arantza Etxeberria for helping clarify this important point.

13. Hutchins (1995, chapter 8) details a specific instance in which a ship's propulsion system failed unexpectedly at a critical moment.

14. I first heard the example from Aaron Sloman.

15. Compare Hutchins's (1995, p. 169) treatment of ants on a beach.

16. For a review see chapters 1 and 4 of Clark 1989. Also see Cliff 1994. Something close to an explicit endorsement of such rationalistic strategies can be found in Newell and Simon 1981.

17. See section 3.2 above. See also McClelland 1989 and Plunkett and Sinha 1991.

18. See section 3.5 above. See also Kirsh and Maglio 1994.

Chapter 5

1. See e.g. Simon 1962, Dawkins 1986, and chapter 4 of Clark 1989.

2. See chapter 4 of Clark 1989 for a review of this and other possibilities.

3. See e.g. Holland 1975; Goldberg 1989; Koza 1992; Belew 1990; Nolfi, Floreano, Miglino, and Mondada 1994.

4. See chapter 3 above.

5. See introduction and chapter 2 of the present volume.

6. For some simulated evolution experiments involving pursuit and evasion, see Miller and Cliff 1994.

7. In this vein, Menczer and Belew (1994) use a genetic algorithm to determine the choice of an organism-environment interface by evolving different types of sensor.

8. For a lovely treatment of the complexities of gene-environment interactions, see Gifford 1990.

9. Nolfi, Miglino, and Parisi 1994 is one of the few attempts to introduce phenotypic plasticity into a combined genetic algorithm and neural network model. In their model the genotype-to-phenotype mapping is a temporally extended and environmentally sensitive process. In addition, evolutionary search is itself used to determine the balance between the influence of the genes and of the environment.

10. See e.g. Brooks 1992 and Smithers 1994.

11. Recall the example of the factory assembly robot in the preface.

12. These drawbacks are identified and discussed on p. 194 of Nolfi, Floreano, Miglino, and Mondada 1994.

13. This fine tuning may be achieved by continued evolution, using real robots as the source of genotypes which are then selected and modified using a genetic algorithm simulation. Or it may be achieved by hand tuning and design. For a discussion see Nolfi, Floreano, Miglino, and Mondada 1994.

14. The direct precursors of the new wave of "cognitive dynamicists" were the wonderful cyberneticists of the 1940s and the early 1950s. Landmark publications include Norbert Wiener's *Cybernetics, or Control and Communication in the Animal and in the Machine* (Wiley, 1948), various volumes reprinting (in verbatim transcribed discussions) the goings-on at a series of Macy conferences on cybernetics (Transactions of the Sixth, Seventh, Eighth and Ninth (1949–1952) Macy Conferences (Josiah Macy Jr. Foundation)), W. Ross Ashby's *Introduction to Cybernetics* (Wiley, 1956), and Ashby's classic *Design for a Brain* (Chapman and Hall, 1952).

15. This was the controller that received constant sensory feedback—Beer calls it the "reflexive controller."

Chapter 6

1. I owe this usage to Josefa Toribio.

2. This is also known (less transparently, I feel) as "homuncular explanation"—a usage that reflects the idea that the subcomponents can be miniature intelligent

systems, as long as they, in turn, can be analyzed into smaller and "stupider" parts. The bottom line, of course, is a collection of parts so dumb that they can actually be built. The flip/flops which underlie digital computer circuitry are a nice example of such a physically implementable bottom line. See e.g. Dennett 1978a.

3. See e.g. Bechtel and Richardson 1992.

4. For an argument that emergence is best treated as species of reduction, see Wimsatt 1986 and Wimsatt (to appear).

5. See e.g. Newell and Simon 1976 and Haugeland 1981.

6. This point is made by Kelso (1995, p. 9) and by Ashby (1956). Ashby (ibid., p. 54) states that "the concept of 'feedback,' so simple and natural in certain elementary cases, becomes artificial and of little use when the interconnections between the parts become more complex. When there are only two parts joined so that each affects the other, the properties of feedback give important and useful information about the properties of the whole. But when the number of parts rise to even as few as four, if every one affects the other three, then twenty circuits can be traced through them; and knowing the properties of all the twenty circuits does not give complete information about the system. Such complex systems cannot be treated as an interlaced set of more or less independent feedback circuits, but only as a whole."

7. Actually, complex feedback and feedforward properties of hi-fi circuitry can yield acoustic characteristics which would properly be described as emergent (see below). But manufacturers usually work hard to reduce such interactions, to simplify the signal passing between components and to insulate them against feedback, nonlinear interactions, and the like. See Wimsatt (to appear) for discussion.

8. A nonlinear relation is one in which the two quantities or values do not alter in smooth mutual lockstep. Instead, the value of one quantity may (e.g.) increase for some time without affecting the other at all, and then suddenly, when some hidden threshold is reached, cause the other to make a sudden leap or change. The evolution equation for complex connectionist systems is usually highly nonlinear because a unit's output is not simply the weighted sum of its inputs but instead involves thresholds, step functions, or other sources of nonlinearity. Multiple nonlinear interactions characterize the very strongest forms of emergence. When the interactions are linear and few in number, it is seldom necessary to define collective variables to help explain system behaviors. (I thank Pete Mandik and Tim Lane for insisting on the importance of complex, nonlinear interactive modulations in fixing the very strongest class of cases.) Typical scientific usage, it should be noted, allows the use of the label "emergent" in a variety of much weaker cases too—hence our attention to the wall-following and pole-centering robots, and to the use of the idea of emergence in connection with the broader class of unprogrammed, uncontrolled or environmentally mediated types of adaptive success. For more discussion, see Wimsatt (to appear).

9. See also Bechtel and Richardson 1992.

10. See Norton 1995 for a good introduction.

11. For a fuller account see Salzman 1995.

12. It is, of course, unproblematic that at some level brain, body, and world all "obey the same principles"—the basic laws of subatomic physics constitute just such a level. It is, nonetheless, clear that this does not constitute the optimal level for understanding many phenomena (e.g. how a car engine works). The real claim hereabouts is that there exist basic principles and laws that govern all complex, nonequilibrium dynamical systems and which constitute an optimal level of analysis for the understanding patterns in both neural and bodily behavior.

13. See pp. 54–61 of Kelso 1995 for details of these results and of the mathematical model employed.

14. The title is "If you can't make one, you don't know how it works." Dretske's view (1994, pp. 468–482) is that, despite some superficial problems, this claim is true "in all the relevant senses of all the relevant words."

15. See also Lichtenstein and Slovic 1971.

16. For more detail, see chapter 8 of Damasio 1994 and the discussion in chapter 7 below.

Chapter 7

1. See Newell and Simon 1972 and pp. 151–170 of Boden 1988.

2. Zenon Pylyshyn, one of the leading theorists in the field, wrote that cognitive science, given a computational slant, allowed for a "study of cognitive activity fully abstracted in principle from both biological and phenomenological foundations . . . a science of structure and function divorced from material substance" (1986, p. 68).

3. See essays in Gluck and Rumelhart 1990 and in Nadel et al. 1989, and several of the contributions in Koch and Davis 1994.

4. Recall the discussion in chapter 3 above.

5. The case of finger-movement control appears to lie at the "highly distributed" end of a continuum of coding possibilities. At the opposite end, we do indeed find some coding schemes that use spatial groupings of neurons to support an inner topographic map (an inner map that preserves spatial relations among sensory inputs). For example, there is a group of neurons in rat cerebral cortex whose spatial organization echoes the spatial layout of the rat's whiskers. Even in such apparently clear-cut cases, however, it is worth noting that the inner topography is keyed to the maximal responses of the individual neurons and thus leaves room for other aspects of the tuning of such neurons to play a role (see section 7.3 below), and that the response profiles are typically obtained in artificial situations (involving the use of electrical or surgical manipulations) and may not faithfully reflect the role of the neurons in responding to ecologically normal situations. Nonetheless, the existence of the inner topographic mapping is a striking and important result which shows that nature may use a number of rather different

strategies and ploys to promote adaptive success. For a discussion of the case of the rat's whiskers, see Woolsey 1990.

6. This discussion draws heavily on Van Essen and Gallant 1994.

7. See especially Van Essen and Gallant 1994, Knierim and Van Essen 1992, and Felleman and Van Essen 1991. Much of the work described is actually based on studies of the macaque monkey, whose visual system seems usefully similar to the human one.

8. "Differences in firing rate convey information useful for discriminating among stimuli that lie on the slopes of each cell's multidimensional tuning surface." (Van Essen and Gallant 1994, p. 4)

9. Posner (1994) points out this distinction.

Chapter 8

1. This is defined below.

2. See introduction and chapter 3 above, Smolensky 1988, Fodor and Pylyshyn 1988, and Clark 1989.

3. For lots more on state-space encoding see Churchland 1989, Clark 1989, and Clark 1993. For a discussion of the contrast with classical combinational schemes see van Gelder 1990. For a discussion of the special nature of connectionist representational systems see Clark 1994.

4. Nothing here turns on whether or not, in fact, the rat's posterior parietal neurons can act in the absence of the visual inputs (e.g. during dreaming, if rats dream). The point is rather that a lack of decouplability does not in itself seem to deprive the representational gloss of all explanatory force.

5. The most fully worked-out version of this kind of consumption-oriented approach is probably that of Millikan (1994).

6. For details see pp. 49–50 of McNaughton and Nadel 1990.

7. See e.g. Gibson 1979.

8. Consider this smattering of quotes: "Our commitment to a biologically consistent theory means that we categorically reject machine analogies of cognition and development. . . . We deliberately eschew the machine vocabulary of processing devices, programs, storage units, schemata, modules or wiring diagrams. We substitute . . . a vocabulary suited to a fluid, organic system with certain thermodynamic properties. (Thelen and Smith 1994) "We posit that development happens because of the time-locked pattern of activity across heterogeneous components. We are not building representations at all! Mind is activity in . . . the real time of real physical causes." (ibid.) "Representation is the wrong unit of abstraction in building the bulkiest parts of intelligent systems." (Brooks 1991) "The concept of 'representation' . . . is unnecessary as a keystone for explaining the brain and behavior." (Skarda and Freeman 1987) "Explanations in terms of structure in the head—'beliefs,' 'rules,' 'concepts' and 'schemata'—are not acceptable. . . . Our theory has new concepts at the center—nonlinearity, reentrance, coupling het-

erochronicity, attractors, momentum, state spaces, intrinsic dynamics, forces. These concepts are not reducible to the old ones." (Thelen and Smith 1994)

9. The distinction here is between strategies we employ ("on line") to promote quick identification and response during daily action and strategies we have available ("off line") as more reflective, time-consuming, backup procedures. Thus, we may use easy cues like gray hair and beards for daily grandfather detection (or fins and swimming to pick out fish) while being in possession of much more accurate strategies which we could deploy if given more time and information. "On line" thus signifies time-and-resource-constrained daily problem solving—a mode that will favor quick, dirty, semi-automated strategies over more intensive procedures that would reflect our deeper knowledge and commitments.

10. See section 2.7 above. A similar suggestion is made in Hooker et al. 1992, a lovely paper that neatly contrasts various understandings of internal representation and defends a conception of representation as control. See also Clark 1995.

11. For the evolutionary claim see Millikan 1995. For the developmental claim see Karmiloff-Smith 1979, Karmiloff-Smith 1992, and Clark and Karmiloff-Smith 1993. The tempting image of a gradual transition from action-oriented to more action-neutral encodings is discussed briefly in Clark 1995.

12. See Thelen and Smith 1994, pp. 8–20 and 263–266.

13. There remain many tricky questions concerning the right way to understand the notion of computation itself. In particular, some would argue that computation, properly so-called, can occur only in systems that use discrete states rather than (e.g.) units with continuous activation levels. This "formalist" view of computation is thus tied to the ideas of digitalness that underlie many classical results in the theory of computability. There is also, however, a more informal notion of computation (with an equally impressive historical pedigree in early work on so-called analog computation) which is tied to the general idea of automated information processing and the transformation of representations. I have this less restrictive notion of computation in mind throughout the present treatment. Notice that on this account the burden of showing that a system is computational reduces to the task of showing that it is engaged in the automated processing and transformation of information. These notions, alas, are also problematic. For more discussion of all these themes see Giunti 1996, Smith 1995, Smith 1996, Hardcastle 1995, and papers in Harnad 1994.

14. For similar claims see pp. 83–85, 161, and 331–338 of Thelen and Smith 1994 and p. 74 of Thelen 1995.

15. For a full discussion of the complex case of partial genetic specification see pp. 116–117 of Dennett 1995.

16. The case of genetic programming is, however, delicate and interesting. And for reasons which are highly germane to our discussion. Does the genome really code for developmental outcomes? In a certain sense, it does not. It is increasingly clear that most traits or characteristics of individuals result from a complex interplay between multiple genes and local environmental conditions. Should we therefore just give up on the idea of "genes for" such and such altogether? The jury is still out, but several theorists have recently suggested that ideas about genetic

encoding and specification can still make good and useful sense despite the fact that any given gene is at best a partial determinant whose ultimate effects depend heavily on the structure of the environment and the presence of other genes. One reason is that talk of what a given gene is "for:" alerts us to a certain type of functional fact: one whose specification is not problematic as long as the other conditions (the rest of the genes, the local environment) are kept fixed. Thus, it is argued, it is safe and proper to say that a gene is a gene for a long neck if "rivals for that gene's spot in the chromosome would lead in the relevant environment (including genetic environment) to a shorter [neck]" (Sterelny 1995, p. 162). Why grace the genetic contribution with the purposive gloss (gene for a long neck)? Because, quite simply, the genetic material exists to control that feature, whereas the local environmental parameters (usually) do not. Sterelny gives as an example snow gum plants, which develop differently in different climates. The differences have adaptive value, and are caused by the combination of local weather conditions (which act as triggers) and genetic influences. But the genome is structured the way it is precisely so as to allow such climatic adaptation, whereas the weather, as we all know, is supremely indifferent to the fates of living things. For more discussion of all these issues see Oyama 1985, Dawkins 1982, Gifford 1990, Gifford 1994, Dennnett 1995, and Sterelny 1995.

17. The concept of synergy is meant to capture the idea of links or couplings which constrain the collective unfolding of a system comprising many parts. Kelso (1995, p. 38) gives the example of the way the front wheels of a car are constrained to turn at the same time—a built-in synergy which certainly simplifies steering. The same concept can be fruitfully applied to the example of inter-hand coordination patterns discussed in chapter 6 above (see Kelso 1995, p. 52).

18. The danger, of course, is that the notion is now too liberal, allowing (e.g.) library index card systems and fax networks to count as computational systems. I agree these are at best marginal cases. The main alternative notion, however, errs in the opposite direction. For this notion (sometimes called the "formalist idea of computation") ties the very idea of computation to ideas about digital encoding and "classical computability." Yet the notion of analog computation has a long and distinguished history and surely ought not to be ruled a contradiction in terms! For useful discussion of all these themes see Harnad 1994, Hardcastle 1995, Smith 1995, and Giunti 1996.

19. For some examples see pp. 119–120 of Churchland and Sejnowski 1992.

20. For an extended discussion see Port et al. 1995; the preceding paragraphs of my chapter owe much to their clear and concise treatment.

21. I borrow this notion of insulation from Butler (to appear). Chapter 4 of that treatment presents a good case against the radical thesis of embodied cognition, though it fails to address the especially challenging class of cases discussed in the present section.

22. Also known as "circular causation"—see e.g. the cybernetics literature cited in the notes to chapter 5. The notion is also prominent in Kelso 1995. I avoid this phrase because it seems to suggest a simple process involving one stage of feedback from output to input. The most interesting cases of continuous reciprocal

causation involve multiple, asynchronous sources of feedback—see Kelso 1995, p. 9; Ashby 1956, p. 54.

23. This is, in fact, exactly the kind of scenario envisioned by Damasio's convergence-zone hypothesis (see section 7.4).

24. See Clark and Toribio 1994.

25. For a fuller defense of this claim see Clark and Thornton 1996.

26. This question emerged in the course of a particularly fruitful conversation with Randy Beer.

27. This vision of mutual enrichment is powerfully pursued in recent work by Melanie Mitchell and Jim Crutchfield—see e.g. Crutchfield and Mitchell 1995 and Mitchell et al.1994.

28. See essays in Port and van Gelder 1995. See also van Gelder 1995, pp. 376–377.

29. The idea here (which may seem initially paradoxical) is that "narrow contents" (Fodor 1986) may at times supervene on states of the agent plus select chunks of the local environment. See Clark and Chalmers 1995.

30. The worry will thus only apply to choices of dynamical vehicle that are not grounded in some detailed, component-level understanding. Connectionist accounts given in terms of trajectories, state spaces, and attractors (see e.g. Elman 1991) will not be affected, since the basic parameters of such stories are already fixed by the properties of basic components. The same will be true of neurally grounded dynamical accounts (see e.g. Jordan et al. 1994).

31. Of course, there was always a large gap between an algorithmic description and any particular implementation. But a major virtue of standard computational approaches was that they at least constrained the algorithmic stories in ways that guaranteed that we could in principle implement them using only the basic resources of a universal Turing machine—see e.g. Newell and Simon 1981. (What an abstract dynamical description thus loses in stepwise mechanistic prescription, it may however make up for in *temporal* force—see van Gelder 1995.)

32. Other especially relevant works include Maturana and Varela 1987, Dreyfus 1979, Winograd and Flores 1986, Kelso 1995, and some of the papers in Boden 1996.

33. For an excellent discussion see chapters 3 and 6 of Dreyfus 1991.

34. Wheeler (1995) displays this tension and offers a principled extension of the Heideggerian notion of the background as a solution.

35. For a wonderful discussion of the common themes that tie Merleau-Ponty's work to ongoing projects in embedded, embodied cognitive science, see Hilditch 1995. Varela et al. 1991 contains a powerful treatment of many of Merleau-Ponty's themes and is explicitly presented (see pp. xv–xvii) as a modern continuation of that research program.

36. For some differences see pp. 43–48 of Hilditch 1995. The parallels and the differences are again discussed on pp. 203–204 of Varela et al. 1991. Varela et al. argue that Gibson was unduly committed to the view that perceptual invariants

were simply out in the world rather than co-constructed by animal and reality. Varela et al. appear to be committed to the contrary view that such invariants depend on the perceptually guided activity of the organism.

37. The idea of internal states as embodying kinds of information that can be computationally cheaply deployed to guide subsequent action is found, in one form or another, in Ballard 1991, Brooks 1991, Mataric 1991, Chapman 1990, Gibson 1979, Gibson 1982, Neisser 1993, and Turvey et al. 1981.

38. A further area of recent research, in which the themes of embodiment also loom large, concerns the way body-based schemas and images inform much more abstract styles of thinking. The key idea here is that the way we conceptualize rarified domains (moral problems, temporal relations, argument structures, etc.) is heavily dependent on a kind of metaphorical extension of basic, bodily-experience-based notions. Although clearly related in spirit, my concern with the role of the body and world has really been rather different, for I focus on ways in which actual environmental structures and physical interventions reconfigure the space of individual neural computations. The focus on bodily metaphors is pursued in Lakoff 1987, in Johnson 1987, and in chapter 11 of Thelen and Smith 1994.

39. See especially the Bittorio example in chapter 8 of Varela et al. 1991.

40. See pp. 172–179 of Varela et al. 1991, where the influence of Merleau-Ponty's ideas about circular causation is explicitly highlighted.

41. I cannot help but suspect that there is some disagreement among Varela, Thompson, and Rosch on this issue, for in places (e.g. pp. 172–179) their argument deliberately stops short of this radical conclusion whereas elsewhere (e.g. chapter 10) it seems to endorse it. The exegesis is, however, a delicate matter, and the suggestion of internal tension is correlatively tentative.

42. See, for example, the debate between Vera and Simon (extreme liberals) and Touretzky and Pomerleau in *Cognitive Science* 18 (1994). Although cast as a debate about inner symbols rather than about internal representations, this exchange displays exactly the clash of institutions remarked in the text. Touretzky and Pomerleau identify as inner symbols only items that are syntactically arbitrary (what matters is not the physical state *per se* but only its conventional role), relatively passive (manipulated by a distinct processor), and able to enter into recursive role-based episodes of combination and recombination. Vera and Simon count as symbols any inner states or signals whose role is to designate or denote. My own view, it should be apparent, lies somewhere midway between these extremes. I agree with Touretzky and Pomerleau that not every signal that passes around a complex system should count as a symbol (or internal representation). But it is sufficient if a signal be capable of acting as a genuine stand-in (controlling responses in the absence of current environment input) and if it forms part of some kind of representational system. But, as was argued in section 8.1, such additional constraints fall well short of demanding classical, concatenative symbol systems of the kind imagined by Touretzky and Pomerleau. More generally, I find it intuitive that the general idea of an internal representational system should not be tied too closely to our experiences with languages, texts, and artificial grammars. These are specific types of scheme whose properties may say more about

the computational profile of conscious human thought that about the overall genus of representational systems. For more discussion, see Kirsh 1991, van Gelder 1990, chapter 6 of Clark 1993, Touretzky and Pomerleau 1994, and Vera and Simon 1994.

43. See Skarda and Freeman 1987, Beer 1995, Thelen and Smith 1994, Elman 1994, and Kelso 1995.

44. The phrase is memorable, but its authorship rather elusive. I had long attributed it to the Soviet neuroscientist A. R. Luria, but I can find no corroborating evidence. Almost all my cognitive scientific colleagues in England and the United States recognize the phrase but draw a blank on the authorship. I therefore leave it is an exercise for the reader.

Chapter 9

1. For further discussion see McClamrock 1995, Thelen and Smith 1994, Rutkowska 1993, Hutchins 1995, Resnick 1994, and Varela et al. 1991. See also the essays in Boden 1996.

2. See Davidson 1986.

3. The choice between these two perspectives is delicate and controversial. It can wait for chapter 10.

4. We should, however, distinguish the conception of reason as embodied and embedded from the important but still insufficiently radical notion of "bounded rationality"—see section 9.3.

5. The point about voters' behavior is forcefully made in Satz and Ferejohn 1994. The point about institutional change and public policy is made in North 1993.

6. I am especially influenced by Satz and Ferejohn 1994 and by Denzau and North 1995.

7. This is not to claim (falsely) that highly scaffolded choices will always conform to the norms of substantive rationality. Such will be the case only if the institutional scaffolding has itself evolved as a result of selective pressure to maximize rewards, and if the economic environment has remained stable or if the original institutional scaffolding itself built-in sufficient adaptability to cope with subsequent change.

8. Compare the Condorcet Jury Theorem, which states that if (among other things) juror choices are *independent* then a majority vote by a jury will be correct more often than an average juror.

9. For more on the interplay between human learning and cultural artifacts, see Norman 1988 and section 10.3 below.

Chapter 10

1. Recent authors who subscribe to some version of such a view of language include Dennett (1991, 1995), Carruthers (to appear), and possibly Gauker (1990,

1992). Carruthers, in particular, distinguishes very carefully between "communicative" and "cognitive" concerns with language (pp. 44 and 52). In section 10.2 I attempt to clarify some of the similarities and differences between these treatments and the view of language as a computational transformer. In a related vein, McClamrock (1995) offers an interesting account of "embedded language" in which he stresses facts about the external (physical and social) context in which language is used. However, McClamrock's discussion (see e.g. ibid. pp. 116–131) is centered on the debate concerning "internalist" vs. "externalist" theories of meaning. Several of McClamrock's observations nonetheless bear directly on my concerns, and I discuss them in section 10.3. The view I develop owes most to Hutchins's (1995) treatment of the role of external media in constructing extended cognitive systems (see also chapters 4 and 9 above).

2. Richard Gregory (1981) discusses the role of artifacts (including scissors) as means of reducing individual computational load and expanding our behavioral horizons. Daniel Dennett (1995, pp. 375–378) has pursued the same theme, depicting a class of animals as "Gregorian" creatures (named after Richard Gregory)—creatures that exploit designed artifacts as amplifiers of intelligence and repositories of achieved knowledge and wisdom. See also Norman 1988.

3. This, as I noted in chapter 3, is a somewhat broader use than is customary. Much of the Soviet-inspired literature treats scaffolding as intrinsically social. I extend the notion to include all cases in which external structures are coopted to aid problem solving.

4. This idea originates, I think, in Dennett's (1991, chapters 7 and 8) powerful discussion of the role of words as a means of self-stimulation. The discussion of this theme continues in chapter 13 of Dennett 1995.

5. A major focus of both Carruthers's and Dennett's treatments is the relation between language and consciousness. I will not discuss these issues here, save to say that my sympathies lie more with Churchland (1995, chapter 10), who depicts basic consciousness as the common property of humans and many nonlinguistic animals. Language fantastically augments the power of human cognition. But it does not, I believe, bring into being the basic apprehensions of pleasure, pain, and the sensory world in which the true mystery of consciousness inheres.

6. See chapter 2 of Carruthers (to appear) for an extensive discussion.

7. See Clark 1993 (pp. 97–98) and Clark and Thornton (to appear) for further discussion of this phenomenon, especially as it arises in connectionist learning.

8. See Bratman 1987 for a full discussion.

9. For a detailed treatment of this case, including Elman's other main way of solving the problem (by restricting early memory), see Clark 1994.

10. The simple case of physically manipulating Scrabble tiles to present new potential word fragments to a pattern-completing brain (see Kirsh 1995 and chapter 3 above) is a micro version of the same strategy.

11. For example, Bechtel (1996, p. 128) comments that "linguistic representations possess features that may not be found in our internal cognitive representations. For example, written records can endure unchanged for extended periods of time,

whereas our internal 'memory' appears to rely on reconstruction, not retrieval of stored records. Moreover, through the various syntactical devices provided by language, relations between pieces of information can be kept straight (e.g., that a tree fell and a person jumped) that might otherwise become confused (e.g., when linked only in an associative structure such as a simple connectionist network)."

12. It is, I believe, a failure to fully appreciate the multiple roles of public language that sometimes leads the neurophilosopher Paul Churchland to dismiss linguaform expression as just a shallow reflection of our "real" knowledge (see e.g. Churchland 1989, p. 18). For discussion see Clark 1996 and section 10.4 below.

13. A particularly stunning example is the large mangrove forest extending north from Key West to the Everglades region known as Ten Thousand Islands. The black mangroves of this region can reach heights of 80 feet (Landi 1982, pp. 361–363).

14. Two very recent treatments that emphasize these themes have been brought to my attention. Jean-Pierre Changeux (a neuroscientist and molecular biologist) and Alain Connes (a mathematician) suggest that self-evaluation is the mark of true intelligence—see Changeux and Connes 1995. Derek Bickerton (a linguist) celebrates "off-line thinking" and notes that no other species seems to isolate problems in their own performance and take pointed action to rectify them—see Bickerton 1995.

15. Annette Karmiloff-Smith stresses the modality-neutral dimensions of public language in her closely related work on representational redescription. On the relative context independence of the signs and symbols of public language see Kirsh 1991 and chapter 6 of Clark 1993.

16. The idea that advanced cognition involves repeated processes in which achieved knowledge and representation is redescribed in new formats (which then support new kinds of cognitive operation and access) is pursued in much more detail in Karmiloff-Smith 1992, Clark 1993, Clark and Karmiloff-Smith 1994, and Dennett 1994. The original hypothesis of representational redescription was developed by Karmiloff-Smith (1979, 1986).

17. See e.g. Bechtel 1996, pp. 125–131; Clark 1996, pp. 120–125.

18. Dennett (1991) explores just such a intermediate territory. I discuss Churchland's downplaying of language in detail in Clark 1996. For examples of such downplaying see p. 18 of Churchland 1989 and pp. 265–270 of Churchland and Churchland 1996.

19. For an extensive discussion of the paper clip see Petroski 1992.

20. In what follows I gloss over some very large debates about animal language in general and chimpanzee language in particular. See chapter 13 of Dennett 1995 and chapter 10 of Churchland 1995 for well-balanced discussions.

21. For critical discussion see Pinker 1994, Christiansen 1994, chapter 10 of Churchland 1995, and chapter 13 of Dennett 1995.

22. Any attempt to argue the maximally strong case that human language learning involves no special-purpose language-acquisition device in the brain must, however, contend with a rich variety of detailed linguistic argument and evidence.

In particular it must address the "poverty of the stimulus" argument (Pinker 1994), which claims that it is simply not possible to acquire the detailed grammatical competence we do on the basis of the training data to which we are exposed, and assuming only unbiased, general mechanisms of learning. Since my claim is only that reverse adaptation my play some role in downgrading the amount of "native endowment" we must posit, I make no attempt to address these issued here. For a detailed defense of the strong claim see Christiansen 1994.

23. In the philosophical literature this question invites two standard replies. Either we go with the intuitive demarcations of skin and skull, or we assume that the question is really about the analysis of meaning and proceed to debate the pros and cons of the (broadly) Putnamesque doctrine that "meanings just ain't in the head" (Putnam 1975). I propose, however, to pursue a third position: that cognitive processes are no respecters of the boundaries of the skin or skull. That is to say, I claim (1) that the intuitive notion of the mind ought to be purged of its internalist leanings and (2) that the reasons for so doing do not depend on the (debatable) role of truth-condition and real-world reference in fixing the meaning of mental or linguistic tokens. For a full discussion see Clark and Chalmers 1995.

24. I owe the following examples to Beth Preston (1995). See also Beck 1980 and Gibson and Ingold 1993.

25. The vestibulo-ocular reflex (VOR), to take just one example from dozens, stabilizes the image of the world on the retina so as to offset head movement (see e.g. Churchland and Sejnowski 1992, pp. 353–365). This operation is, of course, crucial for human vision. And human consciousness apprehends the world in a way that depends on the correct operation of the VOR. But the computational steps performed by the VOR circuitry do not figure among our conscious contents. If the computational transformations on which the VOR depends were sometimes carried out using some external device (a neural version of an iron lung or a kidney machine), the interplay between conscious states and VOR computations could remain unaltered. So whatever role is played by the presence of consciousness (whatever exactly that means) somewhere in the loop, that role cannot *itself* afford grounds for rejecting the characterization of some external data transformations as part of our cognitive processing. Rather, it could do so only if we bite the bullet and reject as cognitive all processes which are not themselves consciously introspectable. (If the VOR strikes you as too low-level to count as an example of a nonconscious but genuinely cognitive process, replace it with one you like better—e.g., the processes of content-addressable recall, or whatever introspectively invisible acumen underlies your ability to know which rule to apply next in a logical derivation.)

26. For a valuable but very different discussion of issues concerning the implications of an embodied, embedded approach for conceptions of the self, see Varela et al. 1991.

27. See Dawkins 1982 for an especially biologically astute treatment of this kind of case.

28. For an extended discussion of this claim see Clark and Chalmers 1995.

Chapter 11

1. The estimate that the dolphin is too weak by "about a factor of seven" originated with the biologist James Gray. As Triantafyllou and Triantafyllou (1995, p. 66) point out, it is not yet possible to rigorously test this estimate. But it certainly does appear that dolphins generate remarkable amounts of propulsion from rather limited resources—hence the intensive research efforts recently dedicated to unraveling the mysteries of fish locomotion. (See Gray 1968, Hoar and Randall 1978, and Wu et al. 1975 as well as the advanced studies by the Triantafyllou brothers.)

2. The 49-inch, eight-segment, anodized aluminum robot tuna shown in plate 3 is being studied in a testing tank at the Massachusetts Institute of Technology. The work is sketched in Triantafyllou and Triantafyllou 1995. Some detailed earlier studies are reported in Triantafyllou et al. 1993 and in Triantafyllou et al. 1994.

3. Here the embodied, embedded vision simply adds further support to the connectionists' long-standing insistence that neural encoding does not take sentential form. For the connectionist versions of the argument see e.g. Churchland 1989 and Clark 1989.

4. The case of inner rehearsal of sentences (and other modeling of external media) is interestingly intermediate. Here we *do* profit by replicating, internally, the very *gross* dynamics of some external medium. But, as we saw in chapter 10, we need not suppose that such replication involves the creation of any wholly new type of computational resource. Instead, we may use familiar kinds of pattern-completing neural networks, but trained using our experiences with manipulating real external formalisms. For more discussion see chapter 10 of Churchland 1995; see also Rumelhart et al. 1986.

5. Dennett (1991) offers a sustained and enthralling meditation on this theme.

6. For some further discussion see Clark and Chalmers 1995.

7. This implication is increasingly recognized in developmental psychology. See the works discussed in chapter 2, especially Thelen and Smith 1994 and Rutkowska 1993.

Epilogue

1. The ideas and themes pursued in this little fantasy owe much to the visions of Paul Churchland, Patricia Churchland, Daniel Dennett, Marvin Minsky, Gilbert Ryle, John Haugeland, and Rodney Brooks. In bringing these themes together I have tried for maximum divergence between agent-level and brain-level facts. I do not mean to claim dogmatically that current neuroscience unequivocally posits quite such a radical divergence. Several of the issues on which I allow the brain to take a stand remain the subject of open neuroscientific debate. (For a taste of the debate, see Churchland and Sejnowski 1992 and Churchland et al. 1994.) In

view of the literary conceit adopted, explicit supporting references seemed out of place, but they would include especially the following: Dennett 1978a; Dennett 1991; Minsky 1985; Churchland 1989; Haugeland 1995; R. Brooks, "Intelligence without representation," *Artificial Intelligence* 41 (1991): 139–159; G. Ryle, *The Concept Of Mind* (Hutchinson, 1949); C. Warrington and R. McCarthy, "Categories of knowledge; further fractionations and an attempted integration," *Brain* 110 (1987): 1273–1296. For my own pursuit of some of these themes, see Clark 1993 and Clark 1995.

2. Or Mary's, or Mariano's, or Pepa's. The choice of the classic male English name is intended only as a gentle reference to old *Readers Digest* articles with titles like "I am John's liver" and "I am Joe's kidney." These articles likewise gave voice to our various inner organs, allowing them to explain their structures, needs, and pathologies directly to the reader.

3. Thanks to Daniel Dennett, Joseph Goguen, Keith Sutherland, Dave Chalmers, and an anonymous referee for support, advice, and suggestions.

Bibliography

Abraham, R., and Shaw, C. 1992. *Dynamics—The Geometry of Behavior*. Addison-Wesley.

Ackley, D., and Littman, D. 1992. Interactions between learning and evolution. In *Artificial Life II*, ed. C. Langton et al. Addison-Wesley.

Adolph, K., Eppler, E., and Gibson, E. 1993. Crawling versus walking: Infants' perception of affordances for locomotion on slopes. *Child Development* 64: 1158–1174.

Agre, P. 1988. The Dynamic Structure of Everyday Life. Ph.D. thesis, Department of Electrical Engineering and Computer Science, Massachusetts Institute of technology.

Agre, P., and Chapman, D. 1990. What are plans for? In *Designing Autonomous Agents*, ed. P. Maes. MIT Press.

Albus, J. 197 1. A theory of cerebellar function. *Mathematical Biosciences* 10: 25–61.

Alexopoulos, C., and Mims, C. 1979. *Introductory Mycology*, third edition. Wiley.

Ashby, R. 1952. *Design for a Brain*. Chapman & Hall.

Ashby, R. 1956. *Introduction to Cybernetics*. Wiley.

Ashworth, J., and Dee, J. 1975. *The Biology of Slime Molds*. Edward Arnold.

Ballard, D. 1991. Animate vision. *Artificial Intelligence* 48: 57–86.

Baum, C. 1993. The Effects of Occupation on Behaviors of Persons with Senile Dementia of the Alzheimer's Type and Their Carers. Doctoral thesis, George Warren Brown School of Social Work, Washington University, St. Louis.

Bechtel, W., and Richardson, R. 1992. *Discovering Complexity: Decomposition and Localization as Scientific Research Strategies*. Princeton University Press.

Beck, B. 1980. *Animal Tool Behavior: The Use and Manufacture of Tools by Animals*. Garland.

Beckers, R., Holland, O., and Deneubourg, J. 1994. From local actions to global tasks: Stigmergy and collective robotics. In *Artificial Life 4*, ed. R. Brooks and P. Maes. MIT Press.

Beer, R. 1995a. Computational and dynamical languages for autonomous agents. In *Mind as Motion*, ed. R. Port and T. van Gelder. MIT Press.

Beer, R. 1995b. A dynamical systems perspective on agent-environment interaction. *Artificial Intelligence* 72: 173–215.

Beer, R., and Chiel, H. 1993. Simulations of cockroach locomotion and escape. In *Biological Neural Networks in Invertebrate Neuroethology and Robotics*, ed. R. Beer et al. Academic Press.

Beer, R., and Gallagher, J. 1992. Evolving dynamical neural networks for adaptive behavior. *Adaptive Behavior* 1: 91–122.

Beer, R., Ritzman, R., and McKenna, T., eds. 1993. *Biological Neural Networks in Invertebrate Neuroethology and Robotics*. Academic Press.

Beer, R., Chiel, H., Quinn, K., Espenschied, S., and Larsson, P. 1992. A distributed neural network architecture for hexapod robot locomotion. *Neural Computation* 4, no. 3: 356–365.

Belew, R. 1990. Evolution, learning, and culture: Computational metaphors for adaptive algorithms. *Complex Systems* 4: 11–49.

Berk, L. 1994. Why children talk to themselves. *Scientific American* 271, no. 5: 78–83.

Berk, L., and Garvin, R. 1984. Development of private speech among low-income Appalachian children. *Developmental Psychology* 20, no. 2: 271–286.

Bickerton, D. 1995. *Language and Human Behavior*. University of Washington Press.

Bivens, J., and Berk, L. 1990. A longitudinal study of the development of elementary school children's private speech. *Merrill-Palmer Quarterly* 36, no. 4: 443–463.

Boden, M. 1988. *Computer Models of Mind*. Cambridge University Press.

Boden, M. 1996. *Oxford Readings in the Philosophy of Artificial Life*. Oxford University Press.

Bratman, M. 1987. *Intentions, Plans and Practical Reason*. Harvard University Press.

Bratman, M., Israel, D., and Pollack, M. 1991. Plans and resource-bounded practical reasoning. In *Philosophy and AI: Essays at the Interface*, ed. R. Cummins and J. Pollock. MIT Press.

Brooks, R. 1991. Intelligence without reason. In *Proceedings of the 12th International Joint Conference on Artificial Intelligence*. Morgan Kauffman.

Brooks, R. 1993. A robot that walks: Emergent behaviors from a carefully evolved network. In *Biological Neural Networks in Invertebrate Neuroethology and Robotics*, ed. R. Beer et al. Academic Press.

Brooks, R. 1994. Coherent behavior from many adaptive processes. In *From Animals to Animats 3*, ed. D. Cliff et al. MIT Press.

Brooks, R., and Maes, P., eds. 1994. *Artificial Life 4*. MIT Press.

Brooks, R., and Stein, L. 1993. Building Brains for Bodies. Memo 1439, Artificial Intelligence Laboratory, Massachusetts Institute of Technology.

Bruner, J. 1968. *Processes in Cognitive Growth: Infancy*. Clark University Press.

Busemeyer, J., and Townsend, J. 1995. Decision field theory. In *Mind as Motion*, ed. R. Port and T. van Gelder. MIT Press.

Butler, K. (to appear). *Internal Affairs: A Critique of Externalism in the Philosophy of Mind.*

Campos, J., Hiatt, S., Ramsey, D., Henderson, C., and Svejda, M. 1978. The emergence of fear on the visual cliff. In *The Development of Affect*, volume 1, ed. M. Lewis and R. Rosenblum. Plenum.

Chambers, D., and Reisberg, D. 1985. Can mental images be ambiguous? *Journal of Experimental Psychology: Human Perception and Performance* 11: 317–328.

Changeux, J.-P., and Connes, A. 1995. *Conversations on Mind, Matter and Mathematics*. Princeton University Press.

Chapman, D. 1990. Vision, Instruction and Action. Technical Report 1204, Artificial Intelligence Laboratory, Massachusetts Institute of Technology.

Christiansen, M. 1994. The Evolution and Acquisition of Language. PNP Research Report, Washington University, St. Louis.

Churchland, P. M. 1989. *A Neurocomputational Perspective*. MIT Press.

Churchland, P. M. 1995. *The Engine of Reason, the Seat of the Soul*. MIT Press.

Churchland, P. S., and Sejnowski, T. 1992. *The Computational Brain*. MIT Press.

Churchland, P. S., Ramachandran, V. S., and Sejnowski, T. J. 1994. A critique of pure vision. In *Large-Scale Neuronal Theories of the Brain*, ed. C. Koch and J. Davis. MIT Press.

Clark, A. 1986. Superman; the image. *Analysis* 46, no. 4: 221–224.

Clark, A. 1987. Being there; why implementation matters to cognitive science. *AI Review* 1, no. 4: 231–244.

Clark, A. 1988a. Superman and the duck/rabbit. *Analysis* 48, no. 1: 54–57.

Clark, A. 1988b. Being there (again); a reply to Connah, Shiels and Wavish. *Artificial Intelligence Review* 66, no. 3: 48–51.

Clark, A. 1989. *Microcognition: Philosophy, Cognitive Science and Parallel Distributed Processing*. MIT Press.

Clark, A. 1993. *Associative Engines: Connectionism, Concepts and Representational Change*. MIT Press.

Clark, A. 1994. Representational trajectories in connectionist learning. *Minds and Machines* 4: 317–332.

Clark, A. 1995. Moving minds: Re-thinking representation in the heat of situated action. In *Philosophical Perspectives 9: AI Connectionism and Philosophical Psychology*, ed. J. Tomberlin.

Clark, A. 1996. Connectionism, moral cognition and collaborative problem solving. In *Mind and Morals: Essays on Ethics and Cognitive Science*, ed. L. May et al. MIT Press.

Clark, A. (to appear) Philosophical foundations. In *Handbook of Perception and Cognition*, volume 14: *Artificial Intelligence and Computational Psychology*, ed. M. Boden. Academic Press.

Clark, A., and Chalmers, D. 1995. The Extended Mind. Philosophy-Neuroscience-Psychology Research Report, Washington University, St. Louis.

Clark, A., and Karmiloff-Smith, A. 1994. The cognizer's innards: A psychological and philosophical perspective on the development of thought. *Mind and Language* 8: 487–519.

Clark, A., and Thornton, C. (to appear) Trading spaces: Connectionism and the limits of learning. *Behavioral and Brain Sciences*.

Clark, A., and Toribio, J. 1995. Doing without representing? *Synthese* 101: 401–431.

Cliff, D. 1994. AI and a-life: Never mind the blocksworld. *AISB Quarterly* 87: 16–21.

Cole, M., Hood, L., and McDermott, R. 1978. Ecological niche picking. In *Memory Observed: Remembering in Natural Contexts*, ed. U. Neisser. Freeman, 1982.

Connell, J. 1989. A Colony Architecture for an Artificial Creature, Technical Report II 5 1, MIT AI Lab.

Cottrell, G. 1991. Extracting features from faces using compression networks. In *Connectionist Models: Proceedings of the 1990 Summer School*, ed. D. Touretzky et al. Morgan Kauffman.

Crutchfield, J., and Mitchell, M. 1995. The evolution of emergent computation. In *Proceedings of the National Academy of Science* 92: 10742–10746.

Culicover, P., and R. Harnish, eds. *Neural Connections, Mental Computation*. MIT Press.

Damasio, A. 1994. *Descartes' Error*. Grosset Putnam.

Damasio, A., and Damasio, H. 1994. Cortical systems for retrieval of concrete knowledge: The convergence zone framework. In *Large-Scale Neuronal Theories of the Brain*, ed. C. Koch and J. Davis. MIT Press.

Damasio, A., Tramel, D., and Damasio, H. 1989. Amnesia caused by herpes simplex encephalitis, infarctions in basal forebrain, Alzheimer's disease and anoxia. In *Handbook of Neuropsychology*, volume 3, ed. F. Boller and J. Grafman. Elsevier.

Damasio, A., Tramel, D., and Damasio, H. 1990. Individuals with sociopathic behavior caused by frontal damage fail to respond autonomically to social stimuli. *Behavioral and Brain Research* 4: 81–94.

Damasio, H., Grabowski, T., Frank, R., Galaburda, A., and Damasio, A. 1994. The return of Phineas Gage: Clues about the brain from the skull of a famous patient. *Science* 264: 1102–1105.

Davidson, C. 1994. Common sense and the computer. *New Scientist* 142, April 2: 30–33.

Davidson, D. 1986. Rational animals. In *Actions and Events*, ed. E. Lepore and B. McLaughlin. Blackwell.

Davies, M., and Stone, T., eds. 1995. *Mental Simulation: Evaluations and Applications*. Blackwell.

Dawkins, R. 1982. *The Extended Phenotype*. Oxford University Press.

Dawkins, R. 1986. *The Blind Watchmaker*. Longman.

Dean, P., Mayhew, J., and Langdon, P. 1994. Learning and maintaining saccadic accuracy: A model of brainstem-cerebellum interactions. *Journal of Cognitive Neuroscience* 6: 117–138.

Dennett, D. 1978a. *Brainstorms*. MIT Press.

Dennett, D. 1978b. Why not the whole iguana? *Behavioral and Brain Sciences* 1: 103–4.

Dennett, D. 1991. *Consciousness Explained*. Little, Brown.

Dennett, D. 1994. Labeling and learning. *Mind and Language* 8: 54–48.

Dennett, D. 1995. *Darwin's Dangerous Idea: Evolution and the Meanings of Life*. Simon and Schuster.

Denzau, A., and North, D. 1995. Shared Mental Models: Ideologies and Institutions. Unpublished.

Diaz, R., and Berk, L., eds. 1992. *Private Speech: From Social Interaction to Self-Regulation*. Erlbaum.

Donald, M. 1991. *Origins of the Modern Mind*. Harvard University Press.

Dretske, F. 1988. *Explaining Behavior: Reasons in a World of Causes*. MIT Press.

Dretske, F. 1994. If you can't make one you don't know how it works. In *Midwest Studies in Philosophy XIX: Philosophical Naturalism*. University of Notre Dame Press.

Dreyfus, H. 1979. *What Computers Can't Do*. Harper & Row.

Dreyfus, H. 1991. *Being-in-the-World: A Commentary on Heidegger's Being and Time, Division I*. MIT Press.

Dreyfus, H., and Dreyfus, S. 1990. What is morality? A phenomenological account of the development of ethical experience. In *Universalism vs. Communitarianism*, ed. D. Rasmussen. MIT Press.

Edelman, G. 1987. *Neural Darwinism*. Basic Books.

Edelman, G., and Mountcastle, V. 1978. *The Mindful Brain*. MIT Press.

Edwards, D., Baum, C., and Morrow-Howell, N. 1994. Home environments of inner city elderly with dementia: Do they facilitate or inhibit function? *Gerontologist* 34, no. 1: 64.

Elman, J. 1991. Representation and structure in connectionist models. In *Cognitive Models of Speech Processing*, ed. G. Altman. MIT Press.

Elman, J. 1994. Learning and development in neural networks: The importance of starting small. *Cognition* 48: 71-99.

Elman, J. 1995. Language as a dynamical system. In *Mind as Motion*, ed. R. Port and T. van Gelder. MIT Press.

Farah, M. 1990. *Visual Agnosia*. MIT Press.

Farr, M. 1981. *How to Know the True Slime Molds*. William Brown.

Feigenbaum, E. 1977. The art of artificial intelligence: 1. Themes and case studies of knowledge engineering. In Proceedings of the Fifth International Joint Conference on Artificial Intelligence.

Felleman, D., and Van Essen, D. 1991. Distributed hierarchial processing in primate visual cortex. *Cerebral Cortex* 1: 1–47.

Fikes, R., and Nilsson, N. 1971. STRIPS: A new approach to the application of theorem proving to problem solving. *Artificial Intelligence* 2: 189–208.

Fodor, J., and Pylyshyn, Z. 1988. Connectionism and cognitive architecture. *Cognition* 28: 3–71.

Friedman, M. 1953. *Essays in Positive Economics*. University of Chicago Press.

Gallistel, C. 1980. *The Organization of Behavior*. Erlbaum.

Gauker, C. 1990. How to learn a language like a chimpanzee. *Philosophical Psychology* 3, no. 1: 31–53.

Gesell, A. 1939. Reciprocal interweaving in neuromotor development. *Journal of Comparative Neurology* 70: 161–180.

Gibson, J. J. 1979. *The Ecological Approach to Visual Perception*. Houghton Mifflin.

Gibson, J. J. 1982. Reasons for realism. In *Selected Essays of James J. Gibson*, ed. E. Reed and R. Jones. Erlbaum.

Gibson, K., and Ingold, T., eds. 1993. *Tools, Language and Cognition in Human Evolution*. Cambridge University Press.

Gifford, F. 1990. Genetic traits. *Biology and Philosophy* 5: 327–347.

Giunti, M. 1996. Is Computationalism the Hard Core of Cognitive Science? Paper presented at Convego Triennale SILFS, Rome.

Gluck, M., and Rumelhart, D., eds. 1990. *Neuroscience and Connectionist Theory*. Erlbaum.

Gode, D., and Sunder, S. 1992. Allocative Efficiency of Markets with Zero Intelligence (ZI) Traders: Markets as aPartial Substitute for Individual Rationality. Working Paper No. 1992-16, Carnegie Mellon Graduate School of Industrial Administration.

Goldberg, D. 1989. *Genetic Algorithms in Search, Optimization, and Machine Learning*. Addison-Wesley.

Goldman, A. 1992. In defense of the simulation theory. *Mind and Language* 7, no. 1–2: 104–119.

Gopalkrishnan, R., Triantafyllou, M. S., Triantafyllou, G. S., and Barrett, D. 1994. Active vorticity control in a shear flow using a flapping foil. *Journal of Fluid Dynamics* 274: 1–21.

Gordon, R. 1992. The simulation theory: Objections and misconceptions. *Mind and Language* 7, no. 1–2: 1–33.

Grasse, P. P. 1959. La Reconstruction du Nid et les Coordinations Inter-Individuelles chez *Bellicositermes Natalensis* et *Cubitermes* sp. La Theorie de la Stigmergie: Essai D'interpretation des Termites Constructeurs. *Insect Societies* 6: 41–83.

Gray, J. 1968. *Animal Locomotion*. Weidenfeld & Nicolson.

Graziano, M., Anderson, R., and Snowden, R. 1994. Tuning of MST neurons to spiral motions. *Journal of Neuroscience* 14: 54–67.

Greene, P.H. 1972. Problems of organization of motor systems. In *Progress in Theoretical Biology*, volume 2, ed. R. Rosen and F. Schnell. Academic Press.

Gregory, R. 1981. *Mind in Science*. Cambridge University Press.

Grifford, F. 1990. Genetic traits. *Biology and Philosophy* 5: 327–347.

Haken, H., Kelso, J., and Bunz, H. 19985. A theoretical model of phase transitions in human hand movements. *Biological Cybernetics* 51: 347–356.

Hallam, J., and Malcolm, C.A. (to appear) Behavior, perception, action and intelligence: the view from situated robots. *Philosophical Transcripts of the Royal Society, London, A.*

Hardcastle, V. (to appear) Computationalism. *Synthese*.

Hare, M., and Elman, J. 1995. Learning and morphological change. *Cognition* 56: 61–98.

Harnad, S. 1994, ed. Special issue on "What is Computation?" *Minds and Machines* 4, no. 4: 377–488.

Harvey, I., Husbands, P., and Cliff, D. 1994. Seeing the light: Artificial evolution, real vision. In *From Animats to Animals 3*, ed. D. Cliff et al. MIT Press.

Haugeland, J. 1981. Semantic engines: An introduction to mind design. In *Mind Design: Philosophy, Psychology, Artificial Intelligence*, ed. J. Haugeland. MIT Press.

Haugeland, J. 1991. Representational genera. In *Philosophy and Connectionist Theory*, ed. W. Ramsey et al. Erlbaum.

Haugeland, J. 1995. Mind embodied and embedded. In *Mind and Cognition*, ed. Y.-H. Houng and J.-C. Ho. Taipei: Academia Sinica.

Hayes, P. 1979. The naive physics manifesto. In *Expert Systems in the Microelectronic Age.*, ed. D. Michie. Edinburgh University Press.

Haugeland, J. 1995. Mind embodied and embedded. In *Mind and Cognition*, ed. Y.-H. Houng and J.-C. Ho. Academia Sinica.

Hebb, D. 1949. *The Organization of Behavior*. Wiley.

Heidegger, M. 1927. *Being and Time* (translation). Harper and Row, 1961.

Hilditch, D. 1995. At the Heart of the World: Maurice Merleau-Ponty's Existential Phenomenology of Perception and the Role of Situated and Bodily Intelligence in Perceptually-Guided Coping. Doctoral thesis, Washington University, St. Louis.

Hofstadter, D. 1985. *Metamagical Themas: Questing for the Essence of Mind and Pattern*. Penguin.

Hogan, N., Bizzi, E., Mussa-Ivaldi, F. A., and Flash, T. 1987. Controlling multijoint motor behavior. *Exercise and Sport Science Reviews* 15: 153–190.

Holland, J. 1975. *Adaptation in Natural and Artificial Systems*. University of Michigan Press.

Hooker, C., Penfold, H., and Evans, R. 1992. Control, connectionism and cognition: Towards a new regulatory paradigm. *British Journal for the Philosophy of Science* 43, no. 4: 517–536.

Hutchins, E. 1995. *Cognition in the Wild*. MIT Press.

Hutchins, E., and Hazelhurst, B. 1991. Learning in the cultural process. In *Artificial Life II*, ed. C. Langton et al. Addison-Wesley.

Jackendoff, R. (to appear) How language helps us think. *Pragmatics and Cognition*.

Jacob, F. 1977. Evolution and tinkering. *Science* 196, no. 4295: 1161–1166.

Jacobs, R., Jordan, M., and Barto, A. 1991. Task decomposition through competition in a modular connectionist architecture: The what and where visual tasks. *Cognitive Science* 15: 219–250.

Jacobs, R., Jordan, M., Nowlan, S., and Hinton, G. 1991. Adaptive mixtures of local experts. *Neural Computation* 3: 79–87.

Jefferson, D. Collins, R., Cooper, C., Dyer, M., Flowers, M., Korf, R., Taylor, C., and Wang, A. 1991. Evolution as a theme in artificial life. In *Proceedings of the Second Conference on Artificial Life*, ed. C. Langton and D. Farmer. Addison-Wesley.

Johnson, M. 1987. *The Body in the Mind: The Bodily Basis of Imagination, Reason and Meaning*. University of Chicago Press.

Johnson, M., Maes, P., and Darrell, T. 1994. Evolving visual routines. In *Artificial Life 4*, ed. R. Brooks and P. Maes. MIT Press.

Jordan, M. 1986. Serial Order: A Parallel Distributed Processing Approach. Report 8604, Institute for Cognitive Science, University of California, San Diego.

Jordan, M., Flash, T., and Arnon, Y. 1994. A model of the learning of arm trajectories from spatial deviations. *Journal of Cognitive Neuroscience* 6, no. 4: 359-376.

Kaelbling, L. 1993. *Learning in Embedded Systems*. MIT Press.

Kagel, J. 1987. Economics according to the rat (and pigeons too). In *Laboratory Experimentation in Economics: Six Points of View*, ed. A. Roth. Cambridge University Press.

Kandel, E., and Schwarz, J. 1985. *Principles of Neural Science*. Elsevier.

Karmiloff-Smith, A. 1979. *A Functional Approach to Child Language*. Cambridge University Press.

Karmiloff-Smith, A. 1986. From meta-process to conscious access. *Cognition* 23: 95–147.

Karmiloff-Smith, A. 1992. *Beyond Modularity: A Developmental Perspective On Cognitive Science*. MIT Press.

Kauffman, S. 1993. *The Origins of Order: Self-Organization and Selection in Evolution*. Oxford University Press.

Kawato, M., et al. 1987. A hierarchical neural network model for the control and learning of voluntary movement. *Biological Cybernetics* 57: 169–185.

Kelso, S. 1995. *Dynamic Patterns*. MIT Press.

Kirsh, D. 1991. When is information explicitly represented? In *Information Thought and Content*, ed. P. Hanson. UBC Press.

Kirsh, D. 1995. The intelligent use of space. *Artificial Intelligence* 72: 1–52.

Kirsh, D., and Maglio, P. 1991. Reaction and Reflection in tetris. Research report D-015, Cognitive Science Department, University of California, San Diego.

Kirsh, D., and Maglio, P. 1994. On distinguishing epistemic from pragmatic action. *Cognitive Science* 18: 513–549.

Kleinrock, L., and Nilsson, A. 1981. On optimal scheduling algorithms for time-shared systems. *Journal of the ACM* 28: 3.

Knierim, J., and Van Essen, D. 1992. Visual cortex: Cartography, connectivity and concurrent processing. *Current Opinion in Neurobiology* 2: 150–155.

Koza, J. 1991. Evolution and co-evolution of computer programs to control independently acting agents. In *From Animals to Animats I*, ed. J.-A. Meyer and S. Wilson. MIT Press.

Koza, J. 1992. *Genetic Programming*. MIT Press.

Laird, J., Newell, A., and Rosenbloom, P. 1987. SOAR: An architecture for general intelligence. *Artificial Intelligence* 33: 1–64.

Lakoff, G. 1987. *Women, Fire and Dangerous Things: What Categories Reveal about the Mind*. University of Chicago Press.

Landi, V. 1982. *The Great American Countryside*. Collier Macmillan.

Le Cun, Y., Boser, B., Denker, J. S., Henderson, D., Howard, R., Hubbard, W., and Jackal, L. 1989. Back propagation applied to handwritten zip code recognition. *Neural Computation* 1: 541–551.

Lenat, D., and Feigenbaum, E. 1992. On the thresholds of knowledge. In *Foundations of Artificial Intelligence*, ed. D. Kirsh. MIT Press.

Lenat, D., and Guha, R. 1990. *Building Large Knowledge-Based Systems: Representation and Inference in the CYC Project*. Addison-Wesley.

Lichtenstein, S., and Slovic, P. 1971. Reversals of preference between bids and choices in gambling decisions. *Journal of Experimental Psychology* 101: 16–20.

Lieberman, P. 1984. *The Biology and Evolution of Language*. Harvard University Press.

Lin, L. 1993. Reinforcement Learning for Robots Using Neural Networks. Doctoral thesis, Carnegie Mellon University.

Mackay, D. 1967. Ways of looking at perception. In *Models for the Perception of Speech and Visual Form*, ed. W. Wathen-Dunn. MIT Press.

Mackay, D. 1973. Visual stability and voluntary eye movements. In *Handbook of Sensory Physiology*, volume VII/3a, ed. R. Jung. Springer-Verlag.

Maes, P. 1994) Modeling adaptive autonomous agents. *Artificial Life* 1, no. 1–2: 135–162.

Magnuson, J. J. 1978. Locomotion by scobrid fishes: Hydromechanics morphology and behavior. In *Fish Physiology*, ed. W. Hoar and D. Randall. Academic Press.

Malone, T., Fikes, R., Grant, K., and Howard, M. 1988. Enterprise: A marker-like task scheduler for distributed computing environments. In *The Ecology of Computation*, ed. B. Huberman. North-Holland.

Marr, D. 1969. A theory of cerebellar cortex. *Journal of Physiology* 202: 437–470.

Marr, D. 1982. *Vision*. Freeman.

Mataric, M. 1991. Navigating with a rat brain: A neurobiologically inspired model for robot spatial representation. In *From Animals to Animats I*, ed. J.-A. Meyer and S. Wilson. MIT Press.

Maturana, H., and Varela, F. 1987. *The Tree of Knowledge: The Biological Roots of Human Understanding*. New Science Library.

McCauley, J. 1994. Finding metrical structure in time. In *Proceedings of the 1993 Connectionist Models Summer School*, ed. M. Mozer et al. Erlbaum.

McCauley, R., ed. 1996. *The Churchlands and Their Critics*. Blackwell.

McClamrock, R. 1995. *Existential Cognition*. University of Chicago Press.

McClelland, J. 1989. Parallel distributed processing—Implications for cognition and development. In *Parallel Distributed Processing: Implications for Psychology and Neurobiology*, ed. R. Morris. Clarendon.

McClelland, J., Rumelhart, D., and Hinton, G. 1986. The appeal of parallel distributed processing. In *Parallel Distributed Processing: Explorations in the Microstructure of Cognition*, volume 1: *Foundations*, ed. D. Rumelhart et al. MIT Press.

McConkie, G. 1979. On the role and control of eye movements in reading. In *Processing of Visible Language*, ed. P. Kolers et al. Plenum.

McConkie, G. 1990. Where Vision and Cognition Meet. Paper presented at HFSP Workshop on Object and Scene Perception, Leuven, Belgium.

McConkie, G., and Rayner, K. 1976. Identifying the span of the effective stimulus in reading: Literature review and theories of reading. In *Theoretical Models and Processes of Reading*, ed. H. Singer and R. Ruddell. International Reading Association.

McCulloch, W., and Pitts, W. 1943. A logical calculus of the ideas immanent in nervous activity. *Bulletin of Mathematical Geophysics* 5: 115–133.

McGraw, M. B. 1945. *The Neuromuscular Maturation of the Human Infant.* Columbia University Press.

McNaughton, B. 1989. Neuronal mechanisms for spatial computation and information storage. In *Neural Connections, Mental Computation*, ed. L. Nadel et al. MIT Press.

McNaughton, B., and Nadel, L. 1990. Hebb-Marr Networks and the neurobiological representation of action in space. In *Neuroscience and Connectionist Theory*, ed. M. Gluck and D. Rumelhart. Erlbaum.

Menczer, F., and Belew, R. 1994. Evolving sensors in environments of controlled complexity. In *Artificial Life 4*, ed. R. Brooks and P. Maes. MIT Press.

Merleau-Ponty, M. 1942. *La Structure du Comportment.* Presses Universites de France. Translation: *The Structure of Behavior* (Beacon, 1963).

Merleau-Ponty, M. 1945. *Phenomenologie de la Perception.* Paris: Gallimard. Translation: *Phenomenology of Perception* (Routledge and Kegan Paul, 1962).

Michie, D., and Johnson, R. 1984. *The Creative Computer.* Penguin.

Miller, G., and Cliff, D. 1994. Protean behavior in dynamic games: Arguments for the co-evolution of pursuit-evasion tactics. In *From Animals to Animats 3*, ed. D. Cliff et al. MIT Press.

Miller, G., and Freyd, J. 1993. *Dynamic Mental Representations of Animate Motion.* Cognitive Science Research Paper 290, University of Sussex.

Millikan, R. 1984. *Language, Thought and Other Biological Categories.* MIT Press.

Millikan, R. 1994. Biosemantics. In *Mental Representation: A Reader*, ed. S. Stich and T. Warfield. Blackwell.

Millikan, R. 1995. Pushmi-Pullyu Representations. In *Philosophical Perspectives 9: AI, Connectionism and Philosophical Psychology*, ed. J. Tomberlin.

Minsky, M. 1985. *The Society of Mind.* Simon & Schuster.

Mitchell, M., Crutchfield, J., and Hraber, P. 1994. Evolving cellular automata to perform computations: Mechanisms and impediments. *Physician D* 75: 361–391.

Morrissey, J. H. 1982. Cell proportioning and pattern formation. In *The Development of Dictyostelium discoideum*, ed. W. Loomis. Academic Press.

Motter, B. 1994. Neural correlates of attentive selection for color or luminance in extrastriate area V4. *Journal of Neuroscience* 14: 2178–2189.

Nadel, L., Cooper, L., Culicover, P., and Harnish, R., eds. 1989. *Neural Connections, Mental Computations.* MIT Press.

Neisser, U. 1993. Without perception there is no knowledge: Implications for artificial intelligence. In *Natural and Artificial Minds*, ed. R. Burton. State University of New York Press.

Newell, A. 1990. *Unified Theories of Cognition*. Harvard University Press.

Newell, A., and Simon H. 1972. *Human Problem Solving*. Prentice-Hall.

Newell, A., and Simon, H. 1981. Computer science as empirical inquiry. In *Mind Design*, ed. J. Haugeland. MIT Press.

Newport, E. 1990. Maturational constraints on language learning. *Cognitive Science* 14: 11–28.

Nolfi, S., Floreano, D., Miglino, O., and Mondada, F. 1994. How to evolve autonomous robots: different approaches in evolutionary robotics. In *Artificial Life 4*, ed. R. Brooks and P. Maes. MIT Press.

Nolfi, S., Miglino, O., and Parisi, D. 1994. Phenotypic Plasticity in Evolving Neural Networks. Technical Report PCIA-94-05, CNR Institute of Psychology, Rome.

Nolfi, S., and Parisi, D. 1991. Auto-Teaching: Networks That Develop Their Own Teaching Input. Technical Report PC1A91-03, CNR Institute of Psychology, Rome.

Norman, D. 1988. *The Psychology of Everyday Things*. Basic Books.

North, D. 1993. Economic Performance Through Time. Text of Prize Lecture in Economic Science in Memory of Alfred Nobel.

Norton, A. 1995. Dynamics: An introduction. In *Mind as Motion*, ed. R. Port and T. van Gelder. MIT Press.

O'Keefe, J. 1989. Computations the hippocampus might perform. In *Neural Connections, Mental Computations*, ed. L. Nadel et al. MIT Press.

O'Regan, J. 1990. Eye movements and reading. In *Eye Movements and Their Role in Visual and Cognitive Processes*, ed. E. Kowler. Elsevier.

O'Regan, K. 1992. Solving the "real" mysteries of visual perception: The world as an outside memory. *Canadian Journal of Psychology* 46: 461–488.

Oyama, S. 1985. *The Ontogeny of Information: Developmental Systems and Evolution*. Cambridge University Press.

Pearson, K. 1985. Are there central pattern generators for walking and flight in insects? In *Feedback and Motor Control in Invertebrates and Vertebrates*, ed. W. Barnes and M. Gladden. Croom Helm.

Petroski, H. 1992. The evolution of artifacts. *American Scientist* 80: 416–420.

Piaget, J. 1952. *The Origins of Intelligence in Children*. International University Press.

Piaget, J. 1976. *The Grasp of Consciousness: Action and Concept in the Young Child*. Harvard University Press.

Pinker, S. 1994. *The Language Instinct*. Morrow.

Plunkett, K., and Sinha, C. 1991. Connectionism and developmental theory. *Psykologisk Skriftserie Aarhus* 16: 1–34.

Polit, A., and Bizzi, E. 1978. Processes controlling arm movements in monkeys. *Science* 201: 1235–1237.

Port, R., Cummins, F., and McCauley, J. 1995. Naive time, temporal patterns and human audition. In *Mind as Motion*, ed. R. Port and T. van Gelder. MIT Press.

Port, R., and van Gelder, T. 1995. *Mind as Motion: Explorations in the Dynamics of Cognition*. MIT Press.

Posner, M., and Rothbart, M. 1994. Constructing neuronal theories of mind. In *Large-Scale Neuronal Theories of the Brain*, ed. C. Koch and J. Davis. MIT Press.

Preston, B. 1995. Cognition and Tool Use (draft paper).

Putnam, H. 1975. The meaning of "meaning." In *Mind, Language and Reality*, ed. H. Putnam. Cambridge University Press.

Pylyshyn, Z. 1986. *Computation and Cognition*. MIT Press.

Quinn, R., and Espenschied, K. 1993. Control of a hexapod robot using a biologically inspired neural network. In *Biological Neural Networks in Invertebrate Neuroethology and Robotics*, ed. R. Beer et al. Academic Press.

Rayner, K, Well, A., and Pollarsek, A. 1980. Asymmetry of the effective visual field in reading. *Perceptual Psychophysics* 27: 537–544.

Resnick, M. 1994a. Learning about life. *Artificial Life* 1, no. _: 229–242.

Resnick, M. 1994b. *Turtles, Termites, and Traffic Jams: Explorations in Massively Parallel Microworlds*. MIT Press.

Ritzmann, R. 1993. The neural organization of cockroach escape and its role in context-dependent orientation. In *Biological Neural Networks in Invertebrate Neuroethology and Robotics*, ed. R. Beer et al. Academic Press.

Rosenblatt, F. 1962. *Principles of Neurodynamics*. Spartan Books.

Rovee-Collier, C. 1990. The "memory system" of prelinguistic infants. In *The Development and Neural Bases of Higher Cognitive Functions*, ed. A. Diamond. New York Academy of Sciences.

Rumelhart, D., and J. McClelland 1986. On learning the past tenses of English verbs. In *Parallel Distributed Processing: Explorations in the Microstructure of Cognition*, volume 2: *Psychological and Biological Models*, ed. J. McClelland et al. MIT Press.

Rumelhart, D., Smolensky, P., McClelland, J., and Hinton, G. 1986. Schemata and sequential thought processes in PDP models. In *Parallel Distributed Processing: Explorations in the Microstructure of Cognition*, ed. D. Rumelhart et al. MIT Press.

Rutkowska, J. 1984. Explaining Infant Perception: Insights from Artificial Intelligence. Cognitive Studies Research Paper 005, University of Sussex.

Rutkowska, J. 1986. Developmental psychology's contribution to cognitive science. In *Artificial Intelligence Society*, ed. K. Gill. Wiley.

Rutkowska, J. 1993. *The Computational Infant*. Harvester Wheatsheaf.

Saito, F., and Fukuda, T. 1994. Two link robot brachiation with connectionist Q-learning. In *From Animals to Animats 3*, ed. D. Cliff et al. MIT Press.

Saltzman, E. 1995. Dynamics and coordinate systems in skilled sensorimotor activity. In *Mind as Motion*, ed. R. Port and T. van Gelder. MIT Press.

Salzman, C., and Newsome, W. 1994. Neural mechanisms for forming a perceptual decision. *Science* 264: 231–237.

Satz, D., and Ferejohn, J. 1994. Rational choice and social theory. *Journal of Philosophy* 9102: 71-87.

Schieber, M. 1990. How might the motor cortex individuate movements? *Trends in Neuroscience* 13, no. 11: 440–444.

Schieber, M., and Hibbard, L. 1993. How somatotopic is the motor cortex hand area? *Science* 261: 489–492.

Shields, P. J., and Rovee-Collier, C. 1992. Long-term memory for context-specific category information at six months. *Child Development* 63: 245–259.

Shortliffe, E. 1976. *Computer Based Medical Consultations: MYCIN*. Elsevier.

Simon, H. 1969. The architecture of complexity. In *The Sciences of the Artificial*, ed. H. Simon. Cambridge University Press.

Simon, H. 1982. *Models of Bounded Rationality*, volumes 1 and 2. MIT Press.

Skarda, C., and Freeman, W. 1987. How brains make chaos in order to make sense of the world. *Behavioral and Brain Sciences* 10: 161–195.

Smith, B. C. 1995. The Foundations of Computation. Paper presented to AISB-95 Workshop on the Foundations of Cognitive Science, University of Sheffield.

Smith, B. C. 1996. *On the Origin of Objects*. MIT Press.

Smithers, T. 1994. Why better robots make it harder. In eds., *From Animals to Animats 3*, ed. D. Cliff et al. MIT Press.

Smolensky, P. 1988. On the proper treatment of connectionism. *Behavioral and Brain Sciences* 11: 1–74.

Steels, L. 1994. The artificial life roots of artificial intelligence. *Artificial Life* 1, no. 1–2: 75–110.

Stein, B., and Meredith, M. 1993. *The Merging of the Senses*. MIT Press.

Sterelny, K. 1995. Understanding life: Recent work in philosophy of biology. *British Journal for the Philosophy of Science* 46, no. 2: 55–183.

Suchman, A. 1987. *Plans and Situated Actions*. Cambridge University Press.

Sutton, R. 1991. Reinforcement learning architecture for animats. In *From Animals to Animats I*, ed. J.-A. Meyer and S. Wilson. MIT Press.

Tate, A. 1985. A review of knowledge based planning techniques. *Knowledge Engineering Review* 1: 4–17.

Thach, W., Goodkin, H., and Keating, J. 1992. The cerebellum and the adaptive coordination of movement. *Annual Review of Neuroscience* 15: 403–442.

Thelen, E. 1986. Treadmill-elicited stepping in seven-month-old infants. *Child Development* 57: 1498–1506.

Thelen, E. 1995. Time-scale dynamics and the development of an embodied cognition. In *Mind as Motion*, ed. R. Port and T. van Gelder. MIT Press.

Thelen, E., Fisher, D. M., Ridley-Johnson, R., and Griffin, N. 1982. The effects of body build and arousal on newborn infant stepping. *Development Psychobiology* 15: 447–453.

Thelen, E. Fisher, D. M., and Ridley-Johnson, R. 1984. The relationship between physical growth and a newborn reflex. *Infant Behavior and Development* 7: 479–493.

Thelen, E., and Smith, L. 1994. *A Dynamic Systems Approach to the Development of Cognition and Action*. MIT Press.

Thelen, E., and Ulrich, B. 1991. Hidden skills: A dynamic system analysis of treadmill stepping during the first year. *Monographs of the Society for Research in Child Development*, no. 223.

Thelen, E., Ulrich, B., and Niles, D. 1987. Bilateral coordination in human infants: Stepping on a split-belt treadmill. *Journal of Experimental Psychology: Human Perception and Performance* 13: 405–410.

Tomasello, Kruger, and Ratner. 1993. Cultural learning. *Behavioral and Brain Sciences* 16: 495–552.

Torras, C. 1985. *Temporal-Pattern Learning in Neural Models*. Springer-Verlag.

Touretzky, D., and Pomerleau, D. 1994. Reconstructing physical symbol systems. *Cognitive Science* 18: 345–353.

Triantafyllou, G. S., Triantafyllou, M. S., and Grosenbaugh, M. A. 1993. Optimal thrust development in oscillating foils with application to fish propulsion. *Journal of Fluids and Structures* 7, no. 2: 205–224.

Triantafyllou, M., and Triantafyllou, G. 1995. An efficient swimming machine. *Scientific American* 272, no. 3: 64–71.

Turvey, M., Shaw, R., Reed, E., and Mace, W. 1981. Ecological laws of perceiving and acting. *Cognition* 9: 237–304.

Valsiner, A. 1987. *Culture and the Development of Children's Action*. Wiley.

Van Essen, D., Anderson, C., and Olshausen, B. 1994. Dynamic routing strategies in sensory, motor, and cognitive processing. In *Large-Scale Neuronal Theories of the Brain*, ed. C. Koch and J. Davis. MIT Press.

Van Essen, D., and Gallant, J. 1994. Neural mechanisms of form and motion processing in the primate visual system. *Neuron* 13: 1–10.

van Gelder, T. 1990. Compositionality: A connectionist variation on a classical theme. *Cognitive Science* 14: 355–384.

van Gelder, T. 1991. Connectionism and dynamical explanation. In Proceedings of the 13th Annual Conference of the Cognitive Science Society, Chicago.

van Gelder, T. 1995. What might cognition be, if not computation? *Journal of Philosophy* 92, no. 7: 345–381.

Varela, F., Thompson E., and Rosch, E. 1991. *The Embodied Mind: Cognitive Science and Human Experience*. MIT Press.

Vera, A., and Simon, H. 1994. Reply to Touretzky and Pomerleau: Reconstructing physical symbol systems. *Cognitive Science* 18: 355–360.

Vogel, S. 1981. Behavior and the physical world of an animal. In *Perspectives in Ethology*, volume 4, ed. P. Bateson and P. Klopfer. Plenum.

Von Foerster, H. 1951., ed. *Cybernetics: Transactions of the Seventh Conference*. Josiah Macy Jr. Foundation.

Von Uexkull, J. 1934. A stroll through the worlds on animals and men. In *Instinctive Behavior*, ed. K. Lashley. International Universities Press.

Vygotsky, L. S. 1986. *Thought and Language* (translation of 1962 edition). MIT Press.

Wason, P. 1968. Reasoning about a rule. *Quarterly Journal of Experimental Psychology* 20: 273–281.

Watkins, C. 1989. Learning from Delayed Rewards. Doctoral thesis, Kings College.

Welch, R. 1978. *Perceptual Modification: Adapting to Altered Sensory Environments*. Academic Press.

Wertsch, J., ed. 1981. *The Concept of Activity in Soviet Psychology*. Sharpe.

Wheeler, M. 1994. From activation to activity. *AISB Quarterly* 87: 36–42.

Wheeler, M. 1995. Escaping from the Cartesian mind-set: Heidegger and artificial life. In Lecture Notes in Artificial Intelligence 929, Advances in Artificial Life, Granada, Spain.

Whorf, B. 1956. *Language, Thought and Reality*. Wiley.

Wiener, N. 1948. *Cybernetics, or Control and Communication in the Animal and in the Machine*. Wiley.

Wimsatt, W. 1986. Forms of aggregativity. In *Human Nature and Natural Knowledge*, ed. A. Donagan et al. Reidel.

Wimsatt, W. (to appear) Emergence as non-aggregativity and the biases of reductionisms. In *Natural Contradictions: Perspectives on Ecology and Change*, ed. P. Taylor and J. Haila.

Winograd, T., and Flores, F. 1986. *Understanding Computers and Cognition: A New Foundation*. Ablex.

Woolsey, T. 1990. Peripheral Alteration and Somatosensory Development. In *Development of Sensory Systems in Mammals*, ed. J. Coleman. Wiley.

Wu, T. Y. T., Brokaw, C. J., Brennen, C., eds. 1975. *Swimming and Flying in Nature*, volume 2. Plenum.

Yamuchi, B., and Beer, R. 1994. Integrating reactive, sequential and learning behavior using dynamical neural networks. In *From Animals to Animats 3*, ed. D. Cliff. MIT Press.

Yarbus, A. 1967. *Eye Movements and Vision*. Plenum.

Zelazo, P. R. 1984. The development of walking: New findings and old assumptions. *Journal of Motor Behavior* 15: 99–137.

Zhang, J., and Norman, D. 1994. Representations in distributed cognitive tasks. *Cognitive Science* 18: 87–122.

Index